Reviews of
Environmental Contamination
and Toxicology

VOLUME 224

T0189219

For further volumes:
http://www.springer.com/series/398

Reviews of Environmental Contamination and Toxicology

Editor
David M. Whitacre

VOLUME 224

ISSN 0179-5953
ISBN 978-1-4899-8686-3   ISBN 978-1-4614-5882-1 (eBook)
DOI 10.1007/978-1-4614-5882-1
Springer New York Heidelberg Dordrecht London

Foreword

International concern in scientific, industrial, and governmental communities over traces of xenobiotics in foods and in both abiotic and biotic environments has justified the present triumvirate of specialized publications in this field: comprehensive reviews, rapidly published research papers and progress reports, and archival documentations. These three international publications are integrated and scheduled to provide the coherency essential for nonduplicative and current progress in a field as dynamic and complex as environmental contamination and toxicology. This series is reserved exclusively for the diversified literature on "toxic" chemicals in our food, our feeds, our homes, recreational and working surroundings, our domestic animals, our wildlife, and ourselves. Tremendous efforts worldwide have been mobilized to evaluate the nature, presence, magnitude, fate, and toxicology of the chemicals loosed upon the Earth. Among the sequelae of this broad new emphasis is an undeniable need for an articulated set of authoritative publications, where one can find the latest important world literature produced by these emerging areas of science together with documentation of pertinent ancillary legislation.

Research directors and legislative or administrative advisers do not have the time to scan the escalating number of technical publications that may contain articles important to current responsibility. Rather, these individuals need the background provided by detailed reviews and the assurance that the latest information is made available to them, all with minimal literature searching. Similarly, the scientist assigned or attracted to a new problem is required to glean all literature pertinent to the task, to publish new developments or important new experimental details quickly, to inform others of findings that might alter their own efforts, and eventually to publish all his/her supporting data and conclusions for archival purposes.

In the fields of environmental contamination and toxicology, the sum of these concerns and responsibilities is decisively addressed by the uniform, encompassing, and timely publication format of the Springer triumvirate:

Reviews of Environmental Contamination and Toxicology [Vol. 1 through 97 (1962–1986) as Residue Reviews] for detailed review articles concerned with any aspects of chemical contaminants, including pesticides, in the total environment with toxicological considerations and consequences.

Bulletin of Environmental Contamination and Toxicology (Vol. 1 in 1966) for rapid publication of short reports of significant advances and discoveries in the fields of air, soil, water, and food contamination and pollution as well as methodology and other disciplines concerned with the introduction, presence, and effects of toxicants in the total environment.

Archives of Environmental Contamination and Toxicology (Vol. 1 in 1973) for important complete articles emphasizing and describing original experimental or theoretical research work pertaining to the scientific aspects of chemical contaminants in the environment.

Manuscripts for Reviews and the Archives are in identical formats and are peer reviewed by scientists in the field for adequacy and value; manuscripts for the *Bulletin* are also reviewed, but are published by photo-offset from camera-ready copy to provide the latest results with minimum delay. The individual editors of these three publications comprise the joint Coordinating Board of Editors with referral within the board of manuscripts submitted to one publication but deemed by major emphasis or length more suitable for one of the others.

Coordinating Board of Editors

Preface

The role of *Reviews* is to publish detailed scientific review articles on all aspects ofenvironmental contamination and associated toxicological consequences. Such articlesfacilitate the often complex task of accessing and interpreting cogent scientificdata within the confines of one or more closely related research fields.

In the nearly 50 years since *Reviews of Environmental Contamination andToxicology* (formerly *Residue Reviews*) was first published, the number, scope, andcomplexity of environmental pollution incidents have grown unabated. During thisentire period, the emphasis has been on publishing articles that address the presenceand toxicity of environmental contaminants. New research is published each yearon a myriad of environmental pollution issues facing people worldwide. This fact,and the routine discovery and reporting of new environmental contamination cases,creates an increasingly important function for *Reviews*.

The staggering volume of scientific literature demands remedy by which data canbe synthesized and made available to readers in an abridged form. *Reviews* addressesthis need and provides detailed reviews worldwide to key scientists and science orpolicy administrators, whether employed by government, universities, or the privatesector.

There is a panoply of environmental issues and concerns on which many scientistshave focused their research in past years. The scope of this list is quitebroad, encompassing environmental events globally that affect marine and terrestrialecosystems; biotic and abiotic environments; impacts on plants, humans, andwildlife; and pollutants, both chemical and radioactive; as well as the ravages ofenvironmental disease in virtually all environmental media (soil, water, air). Newor enhanced safety and environmental concerns have emerged in the last decade tobe added to incidents covered by the media, studied by scientists, and addressedby governmental and private institutions. Among these are events so striking thatthey are creating a paradigm shift. Two in particular are at the center of everincreasingmedia as well as scientific attention: bioterrorism and global warming.Unfortunately, these very worrisome issues are now superimposed on the alreadyextensive list of ongoing environmental challenges.

The ultimate role of publishing scientific research is to enhance understandingof the environment in ways that allow the public to be better informed. Theterm "informed public" as used by Thomas Jefferson in the age of enlightenmentconveyed the thought of soundness and good judgment. In the modern sense, being"well informed" has the narrower meaning of having access to sufficient information. Because the public still gets most of its information on science and technologyfrom TV news and reports, the role for scientists as interpreters and brokers of scientificinformation to the public will grow rather than diminish. Environmentalismis the newest global political force, resulting in the emergence of multinational consortiato control pollution and the evolution of the environmental ethic.Will the new-politics of the twenty-first century involve a consortium of technologists and environmentalists,or a progressive confrontation? These matters are of genuine concernto governmental agencies and legislative bodies around the world.

For those who make the decisions about how our planet is managed, there is anongoing need for continual surveillance and intelligent controls to avoid endangeringthe environment, public health, and wildlife. Ensuring safety-in-use of the manychemicals involved in our highly industrialized culture is a dynamic challenge, forthe old, established materials are continually being displaced by newly developedmolecules more acceptable to federal and state regulatory agencies, public healthofficials, and environmentalists.

Reviews publishes synoptic articles designed to treat the presence, fate, and, ifpossible, the safety of xenobiotics in any segment of the environment. These reviewscan be either general or specific, but properly lie in the domains of analytical chemistryand its methodology, biochemistry, human and animal medicine, legislation,pharmacology, physiology, toxicology, and regulation. Certain affairs in food technologyconcerned specifically with pesticide and other food-additive problems mayalso be appropriate.

Because manuscripts are published in the order in which they are received infinal form, it may seem that some important aspects have been neglected at times.However, these apparent omissions are recognized, and pertinent manuscripts arelikely in preparation or planned. The field is so very large and the interests in itare so varied that the editor and the editorial board earnestly solicit authors andsuggestions of underrepresented topics to make this international book series yetmore useful and worthwhile.

Justification for the preparation of any review for this book series is that it dealswith some aspect of the many real problems arising from the presence of foreignchemicals in our surroundings. Thus, manuscripts may encompass case studies fromany country. Food additives, including pesticides, or their metabolites that may persistinto human food and animal feeds are within this scope. Additionally, chemicalcontamination in any manner of air, water, soil, or plant or animal life is within theseobjectives and their purview.

Manuscripts are often contributed by invitation. However, nominations for new topics or topics in areas that are rapidly advancing are welcome. Preliminary communicationwith the editor is recommended before volunteered review manuscriptsare submitted.

Summerfield, NC, USA   David M. Whitacre

Contents

Microbial Interactions in the Arsenic Cycle: Adoptive Strategies and Applications in Environmental Management 1

Umesh Praveen Dhuldhaj, Ishwar Chandra Yadav, Surendra Singh, and Naveen Kumar Sharma

Chemical Behavior of Phthalates Under Abiotic Conditions in Landfills 39

Jingyu Huang, Philip N. Nkrumah, Yi Li, and Gloria Appiah-Sefah

The Environmental and Human Effects of Ptaquiloside-Induced Enzootic Bovine Hematuria: A Tumorous Disease of Cattle 53

Rinku Sharma, Tej K. Bhat, and Om P. Sharma

Methods for Deriving Pesticide Aquatic Life Criteria for Sediments 97

Tessa L. Fojut, Martice E. Vasquez, Anita H. Poulsen, and Ronald S. Tjeerdema

Index 177

Microbial Interactions in the Arsenic Cycle: Adoptive Strategies and Applications in Environmental Management

Umesh Praveen Dhuldhaj, Ishwar Chandra Yadav, Surendra Singh, and Naveen Kumar Sharma

Contents

1   Introduction... 2
2   Environmental Fate of Arsenic 4
3   Microbial Resistance to Arsenic and Microbial Transformations
    in the Arsenic Cycle... 6
    3.1   Arsenic Reduction.. 7
    3.2   Arsenite Oxidation 7
    3.3   Methylation and Demethylation.. 8
    3.4   Mobilization and Immobilization.. 10
    3.5   Microbial Arsenic Uptake and Extrusion 10
4   The Ars Operon and Proteins... 13
    4.1   Microbial Arsenic Sensing 13
    4.2   Ars Operon and Transcriptional Regulation... 14
    4.3   Arsenate Reductases (ArsCs).. 15
    4.4   Arsenic Permeases (ArsBs)... 16
    4.5   Posttransductional Regulation: ArsA and ArsD................................... 17
    4.6   ArsM and ArsH 19
5   Bioremediation of Arsenic-Contaminated Environments............................... 19
6   Arsenic Biosensors and Measurement of Arsenic Bioavailability................. 25
7   Conclusions 27
8   Summary. 27
References... 28

U.P. Dhuldhaj • I.C. Yadav • S. Singh
CAS in Botany, Banaras Hindu University, Varanasi, UP 221005, India

N.K. Sharma (✉)
Department of Botany, Post Graduate College, Ghazipur, UP 233001, India
e-mail: naveengzp@gmail.com

D.M. Whitacre (ed.), *Reviews of Environmental Contamination and Toxicology*,
Reviews of Environmental Contamination and Toxicology 224,
DOI 10.1007/978-1-4614-5882-1_1, © Springer Science+Business Media New York 2013

1 Introduction

The term "arsenic" is derived from the Persian word "*zarnikh*" meaning "yellow orpiment" (As_2S_3) (Rensing and Rosen 2009). Arsenic (As) is a metalloid (Table 1) and has a single naturally occurring isotope As^{75} (Rensing and Rosen 2009). It occurs in four primary oxidation states, viz., arsenate [As(V)], arsenite [As(III)], elemental arsenic [As(0)], and arsenide [As(−III)]. Arsenic compounds have no known biological roles. Formerly, As found applications in medicine in ancient civilizations (Datta et al. 1979). It has also been used in the cosmetic and agriculture industries (insecticide, desiccant, rodenticide, and herbicide). However, in recent times, the element has acquired notoriety for its toxicity to humans. The Agency for Toxic Substances and Disease Registry (ATSDR) has included arsenic in the list of "20 most hazardous substances" (Rensing and Rosen 2009).

Soluble inorganic arsenic is acutely toxic to organisms (Kaise and Fukui 1992). Arsenite [As(III)] is primarily responsible for the biological effects caused by arsenic. Its affinity for protein thiols (i.e., cysteine thiolates) or vicinal sulfhydryl groups makes it highly toxic. In molecules such as lipoic acid and enzyme pyruvate dehydrogenase, it binds strongly to vicinal cysteine pairs and acts as a potent inhibitor of oxidative metabolism (Lin et al. 2006). The binding of arsenite to protein sulfhydryl group could result in membrane degradation and cell death by producing reactive oxygen species (ROS) (Gebel 2000). Arsenite-stimulated ROSs bind to the reduced glutathione (GSH), cause depletion of the GSH pool (Bhattacharjee et al. 2008) and damage proteins, lipids, and DNA molecules (Liu et al. 2001). Arsenite also acts as an endocrine disrupter by binding to hormone receptors and interfering with normal cell signaling (Kaltreider et al. 2001).

Arsenate [As(V)], another commonly reported form of inorganic arsenic, is a toxic analog of inorganic phosphate (Cervantes et al. 1994; Turpeinen 2002). It competes with inorganic phosphate and acts as uncoupler of oxidative phosphorylation. Owing to its similarity to phosphate, it could produce formation of sugar and nucleotide arsenate, which is less stable than phosphate sugar and nucleotide (Rensing and Rosen 2009). Inorganic arsenic is enzymatically methylated in species such as monomethyl arsenic acid (MMAA) and dimethyl arsenic acid (DMAA), which are less toxic than are the inorganic As(III) and As(V) species (Leonard 1991).

Table 1 The general characteristics of arsenic

Symbol	As
Periodic position	Group V; Period—4; Block—P
Atomic number	33
Atomic weight	74.92 g/mol
Element category	Metalloid
Color	Colorless
Occurrence	It ranks 20th in abundance
Oxidation states	+5, −3, +3, 1
Minerals	Arsenopyrite, realgar, iron pyrite
Crystal structure	Trigonal
Commonly found as	Arsenides of gallium, indium, and aluminum

Table 2 Natural levels of arsenic in different environments (WHO 2001)

Environment	Natural level
Open-ocean water	1.2 μg/L
Fresh surface waters (river, lakes, etc.)	<10 μg/L
Ground water (exp. areas with volcanic rocks and sulfide deposits)	1–2 μg/L
Sediment	5–3,000 mg/kg
Soil	Background concentration 1–40 mg/kg (mean value 5 mg/kg)

In human and other species, because of their low solubility and reduced affinity to tissues, methylated species are likely to have relatively less adverse effect to that of inorganic arsenic species. Mammals (including humans) exposed to arsenic have been reported to contain methylated arsenic species such as dimethyl arsenate DMA(V) and monomethyl arsenate MMA(V) in their urine. The gaseous form arsines are very toxic (Buchet and Lauwerys 1981; Leonard 1991). They combine with hemoglobin of the red blood cells, producing severe swelling of the cells and rendering them nonfunctional (Blair et al. 1990).

Atmospheric concentrations of arsenic are different for rural and urban areas, with an average arsenic concentration range in rural air of 0.02–4 μg/m^3 and in urban air, 3–200 μg/m^3. A much higher concentration (>1,000 μg/m^3) has been reported in the vicinity of industrialized areas. In Table 2, we present the natural levels of arsenic known to exist in different environments, as reported by World Health Organization (WHO 2001). Actual environmental levels change as environmental factors change, viz., geographical area, geology of an area, and proximity to human settlements. Amini et al. (2008) have presented the likely scenario of global distribution of arsenic under reducing and oxidizing conditions. The high arsenic levels in groundwater usually occur in arid or semiarid regions that have high groundwater salinity (Welch and Lico 1998).

In most parts of the world, exposure to arsenic results from consuming arsenic-contaminated water. Exposure also results from inhaling arsine gas, arsenic-contaminated dust, or particulates and consuming arsenic-contaminated foods (Wuilloud et al. 2006). The sensitivity that organisms show to arsenic exposure varies, viz., to concentration, type of arsenic species, and exposure duration. In general, inorganic forms are more toxic than are organoarsenicals, and arsenite is more toxic (by 100-fold) than is arsenate. Cumulative exposure increases risk of health impairment manifold. At arsenic-contaminated industrial sites, arsenate, arsenite, arsenic sulfide, elemental arsenic, and arsine gas are the commonly encountered inorganic arsenic species (Luong et al. 2007).

Acute and long-term exposure to arsenic increases the risk of different types of cancer. Inhaled or ingested inorganic arsenic can injure pregnant women and their fetuses, indicating that arsenic can cross the placenta. Arsenic species have also been reported to occur in breast milk, albeit at low levels (i.e., <2 μg/L) (Aschengrau et al. 1989). In animals, exposure to organic arsenic has resulted in low birth weight, fetal malformations, and fetal deaths (www.epa.gov/ttn/atw/hlthef/arsenic.html). The incidence of arsenicosis is much higher in children (Guo et al. 2001; Rahmann et al. 2001;

Guha Muzumdar 2006). Exposure to arsenic in early life (including gestation and early childhood) increases the rate of mortality (Marshall et al. 2007).

The human sensitivity to arsenic varies with the individual, perhaps because of differential expression of genes involved in the synthesis of enzymes AQP9 (the liver isoform) (Acharyya et al. 1999). There also appears to be a relationship between arsenic sensitivity and nutrition status. Expression of AQP9 can be elevated by nutritional restriction (Carbrey et al. 2003). For example, villagers in West Bengal and Bangladesh, the areas most severely affected by As contamination, who received poor diets had enhanced level of AQP9 in the liver and thereby showed higher rates of arsenite uptake. City inhabitants are more tolerant to enhanced AQP9 levels than are rural inhabitants, because they receive better nutrition (Mazumdar 2008). The deficiency of calcium, foliate, animal proteins, and other dietary vitamins also enhances arsenic susceptibility (Mitra et al. 2004). In addition to inducing cancer and reproductive effects, arsenic may also cause arsenicosis, dermatological disorders, neuropathy, cardiovascular and peripheral vascular disease, diabetes mellitus, obstetrics outcomes, and metabolic pathway disorders (Abernathy et al. 2003; Beane Freeman et al. 2004; Ahamed et al. 2006).

In the following sections, we analyze the role of microbes in the arsenic cycle, and their possible use in bioremediation of arsenic-contaminated environments. This is an attractive and eco-friendly approach because many of these microbes are found in heavy-metal-contaminated water and soil, where they variously transform and accumulate the contaminants, thereby reducing their availability. Further, understanding the mechanism(s), genes and proteins employed by microbes to encounter arsenic toxicity will help in developing arsenic-resistant agriculturally important plants.

2   Environmental Fate of Arsenic

Arsenic is widely distributed in the earth's crust, with an average concentration of 2 mg/kg (i.e., 1 ppm). The free metal is uncommon. Arsenic commonly occurs in trace quantities in rocks, soils, water, and air. In soil, arsenic exists primarily as inorganic arsenite and arsenate forms (Cullen and Reimer 1989; Masscheleyn et al. 1991; Pantsar-Kallio and Korpela 2000; Balasoiu et al. 2001). More than 300 arsenic minerals occur in nature, and of these ~60% occur as arsenates, 20% as sulfides and sulfosalts, and 10% as oxides (Drahota and Filippi 2009). The rest exist as arsenites, arsenides, native elements, and metal alloy (Bowell and Parshley 2001).

In primary arsenic-bearing minerals, As is present as anions (arsenide) or dianion (diarsenide) or as sulfarsenide anions bound to metals such as iron, cobalt, and nickel. Arsenopyrite (FeAsS) is the most abundant form of arsenic (Drahota and Filippi 2009). The secondary arsenite minerals are rare in natural environments (Drahota and Filippi 2009). Upon exposure to the atmosphere and water, primary minerals are converted to secondary minerals, such as arsenic oxides, arsenite, and arsenate minerals complexed with mono-, di-, and tri-valent cations. Surface structure and surface complexing of secondary As minerals have important roles in regulating their mobility in natural systems. Drahota and Filippi (2009) have summarized

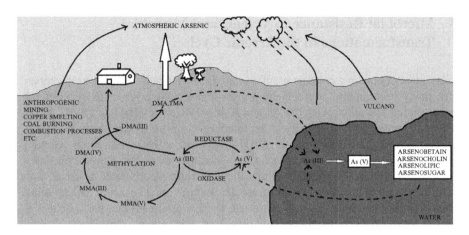

Fig. 1 The global arsenic cycle is depicted. Methylated species such as monomethyl arsenic acid (MMAA), dimethyl arsenic acid (DMAA), and trimethyl arsine oxide (TMAO) dominate arsenic residues in biomass and are also present in soils (after Bhattacharjee and Rosen 2007)

the structure, complexing behavior, and mobility of secondary arsenic minerals in the environment.

Atmospheric flux is a major contributor ($\approx 1/3$) to the arsenic cycle. Volcanic activities and hydrothermal sources (e.g., geysers, fumaroles) are major sources of arsenic pollution, followed by natural low-temperature volatilization of organoarsenic compounds from soils. Anthropogenic activities, such as mining, copper smelting, coal burning, and other combustion processes, contribute arsenic residues to the environment. Arsenic is also produced by the reduction of arsenic trioxide (As_2O_3), which is a byproduct of the copper smelting operations with charcoal. Furthermore, arsenic (arsenopyrite) that remains chemically associated with copper and gold is freed during roasting of the ores (Ng et al. 2003).

Humans contaminate the soil with arsenic from their metalliferous mining and smelting activities, industrial activities, and use of pesticides in agricultural and wood preservatives in the building industry, animal feeds, paints, and dyes. Semiconductors often contain environmentally elusive As residues (Ross 1994; Rensing and Rosen 2009). In the USA, the sodium and calcium salts of monomethylarsenate (MMA) and dimethylarsinate (DMA) are widely used as herbicides and pesticides (Bhattacharjee and Rosen 2007; Rensing and Rosen 2009). Globally, nearly 70% of global arsenic production is used to treat construction timber (as copper chromate arsenate, CCA); 22% is used in agricultural chemicals, and the rest in glass, pharmaceuticals, and nonferrous alloy industries (WHO 2001). However, the United States Environmental Protection Agency (EPA) banned the use of CCA in the USA in the mid-2000s.

Natural forms of arsenic are also important sources of direct contamination of water. Inorganic arsenic of geological origin reaches ground water, which serves as the main source of drinking water in many parts of the world. The solubility of arsenic in water can be substantial and depends on the pH and ionic composition of the water. In Fig. 1, we depict the arsenic cycle that operates in global ecosystems.

3 Microbial Resistance to Arsenic and Microbial Transformations in the Arsenic Cycle

Like other organisms, arsenic at sufficient exposure levels is lethal to microbes. Yet, many bacteria and phytoplankton grow and survive arsenic toxicity through transforming arsenic species (Silver and Keach 1982; Oremland and Stolz 2005). There are several microbe species that tolerate very high level of arsenic toxicity (Table 3). Wolfe-Simon et al. (2011) isolated a bacterial strain GFAJ-1 (Halomonadaceae) from Mono Lake, CA, which uses arsenic in place of phosphorous for its growth. They reported that the bacterium incorporated arsenate, as a substitute for phosphate in macromolecules such as nucleic acids and proteins. However, this study was heavily criticized, and it was argued that the conclusions drawn were not supported by the data presented. For example, any such organism will have functional arsenic-containing enzymes, which were not confirmed to exist in the study (Pennisi 2010).

Microbes transform arsenic metals via redox conversion (a detoxification process involving reductases and oxidases) of inorganic forms [e.g., $As(III) \leftrightarrow As(V)$]

Table 3 Examples of arsenic-resistant microbes

Organism	Reported maximum arsenic tolerance level	References
Escherichia coli	250 mM	Xu et al. 1996; Anderson and Cook 2004
Stenotrophomonas maltophilia SA Ant 15	10 mM As(III), 20 mM As(V)	Botes et al. 2007
Serratia marcescens	15 mM As(III), 500 mM As(V)	Botes et al. 2007
Pseudomonas putida strains RS-5	66.7 mM arsenate	Chang et al. 2008
Pseudomonas putida strains RS-4	66.7 mM arsenate	Chang et al. 2008
Pseudomonas spp., *Acinetobacter* sp.	1,000 mM	Nagvenkar and Ramaiah 2010
Betaproteobacteria or Flavobacteria	100 mM	Jackson et al. 2005
Exiguobacterium, *Aeromonas* spp.	100 mM As(V), 20 mM As(III)	Anderson and Cook 2004
Bacillus sp., *Bacillus licheniformis*	10 mM	Anderson and Cook 2004; Clausen 2004; Bhat 2007
Leptospirillum ferriphilum	3 mM	Tuffin et al. 2006
Acidithiobacillus caldus	20 mM As(V), 30 mM As(III)	Kotze et al. 2006
Streptomyces sp. strain FR-008	100 mM As(V), 5 mM As(III)	Wang et al. 2006
Agrobacterium sp.	25 mM	Cai et al. 2009
Arthrobacter	20 mM	Cai et al. 2009
Rhodococcus	20 mM	Cai et al. 2009
Stenotrophomonas spp.	20 mM	Cai et al. 2009
Acidithiobacillus ferrooxidans	20 mM As(III)	Dave et al. 2008
Corynebacterium glutamicum ATCC 13032	300 mM As(V), 10 mM As(III)	Ordonez et al. 2005
Bacterium GFAJ-1	5 mM	Wolfe-Simon et al. 2011

and mutual conversion of inorganic forms to organic forms (e.g., methylation and demethylation), and vice versa. Microbial redox processes transform highly toxic arsenite [As(III)] to less harmful arsenate form [As(V)]. It was Green (1918) who first identified the microbe-mediated arsenic metabolism (arsenate reducer *Bacterium arsenoreducens* and an arsenite oxidizer *Bacillus arsenoxydans*). Interestingly, a strain of *Thermus* exhibited both As(III) oxidation and As(V) reduction mechanisms (Gihring and Banfield 2001).

3.1 Arsenic Reduction

There are number of bacterial species (e.g., *Sulfurospirillum barnesii*, *Desulfotomaculum auripigmentum*, *Bacillus arsenicoselenatis*, *Chrysiogenes arsenatis*, *Sphingomonas*, *Pseudomonas*, and *Wolinella* spp.) that are capable of reducing As(V) to As(III). Though taxonomically diverse, many microbes utilize (metabolically versatile) arsenic to fulfill their energy demand (Oremland and Stolz 2003). Bacteria categorized as DARPs (dissimilatory arsenate-respiring prokaryotes) use As(V) as an electron acceptor during anaerobic respiration. They oxidize a range of organic (e.g., lactate, acetate, formate, and aromatics) or inorganic substrates (hydrogen and sulfide) as electron donors, resulting in the formation of As(III). The reaction is catalyzed by the enzyme arsenate reductase (Arr) (Fig. 2). Many microbes reduce As(V) to As(III) to survive high environmental arsenic concentrations (i.e., arsenate resistance microbes, ARMs), without obtaining energy from the process (Oremland and Stolz 2005).

From a thermodynamic view point, reduction of arsenate is energetically favorable and provides enough energy for microbial growth (Laverman et al. 1995). However, the overall toxicity of arsenic is likely to restrict the distribution of this process among bacteria (Jackson et al. 2003). In Table 4, we show two pathways by which arsenic can be reduced.

3.2 Arsenite Oxidation

Bacteria such as *Pseudomonas arsenitoxidans*, *Thermus aquaticus*, and *Thermus thermophilus* oxidize As(III) to As(V) (Ilyaletdinov and Abdrashitova 1981; Gihring et al. 2001). CAO (chemoautotrophic arsenite oxidizer, use CO_2/HCO_3^- as carbon source) bacteria oxidize arsenite to arsenate using oxygen, nitrate, or Fe^{3+} as terminal electron acceptor. In contrast, HAOs (heterotrophic arsenite oxidizers) require organic carbon as energy and carbon source (Oremland and Stolz 2005). Arsenite oxidation is an exergonic process (i.e., release of energy) catalyzed by periplasmic arsenite oxidase (aox) (Oremland and Stolz 2003) (Fig. 2). This enzyme has been reported to occur both in heterotrophic and chemotrophic bacteria (Jackson et al. 2003). In *Agrobacterium tumefaciens*, the arsenite oxidase gene contains sensor

Table 4 Arsenite reduction pathways

Reduction pathways		
A five-step mechanism		A multistep mechanism
(ArsC-catalyzed arsenate reduction, based on biochemical studies and X-ray crystallographic structures of free and ligand bound forms of ArsC) (Martin et al. 2001; DeMel et al. 2004)		pI258 ArsC-catalyzed arsenate reduction: based on the structure of the reduced and oxidized forms of pI258 ArsC (NMR and kinetics data) (Zegers et al. 2001; Roos et al. 2006)
Step 1	Nucleophilic attack by Cys12 on non-covalently bound arsenate at the active site, resulting in the formation of a thioarsenate binary product and release of OH^-	Nucleophilic displacement reaction by Cys-10 arsens adduct (Zegers et al. 2001), which is equivalent of the Cys-12 arsens adduct of R773 ArsC (Martin et al. 2001)
Step 2	Nucleophilic attack on the arsenate by GSH results in the formation of {ArsC-Cys12} S-As(IV)-S {glutathione} tertiary complex, H_2O is release (Liu and Rosen 1997)	Arsenite is released after a nucleophilic attack by Cys82 on the covalent Cys10-arsens adduct. An oxidized Cys10–Cys82 intermediate is formed (Messens et al. 2002)
Step 3	Binding of Grx, reduction of arsenate to the dihydroxy monothiol As(III) intermediate, release of OH^-, and a mixed disulfide complex of glutathione and glutaredoxin	Cys89 attacks the Cys10–Cys82 disulfide, resulting in the formation of oxidized Cys82–Cys89 disulfide and reduction of the Cys10 (Zegers et al. 2001; Messens et al. 2002)
Step 4	Formation of a monohydroxy positively charged As(III), with release of OH^-	
Final Step	Addition of OH^- releases free arsenite [As(OH)$_3$] and regenerates free enzyme	ArsC is regenerated by thioredoxin that reduces the Cys82–Cys89 disulfide

kinase system *aoxS* (sensor) and *aoxR* (response regulator). It has two subunits encoded by the genes *aoxA/aroB/asoB* and *aoxB/aroA/asoA*, respectively (Silver and Phung 2005). The *aoxB* genes are specific for the arsenite-oxidizing bacteria. Strains containing both arsenite oxidase gene (*aoxB*) and arsenite transporter gene (*ACR3* or *arsB*) showed enhanced arsenite resistance than those possessing arsenite transporter genes only (Cai et al. 2009). The enzyme ArsAB, a dimmer of ArsA and ArsB, is located in the periplasm. Under anaerobic condition, the *arsAB* genes are expressed at nanomolar arsenate or arsenite concentrations (Rensing and Rosen 2009).

3.3   Methylation and Demethylation

The methylation of As takes place via reduction of As(V) to As (III) and subsequent reduction and addition of a methyl groups (Challenger 1945). McBride and Wolfe (1971) were the first to describe the conversion of As(V) to small amounts of volatile

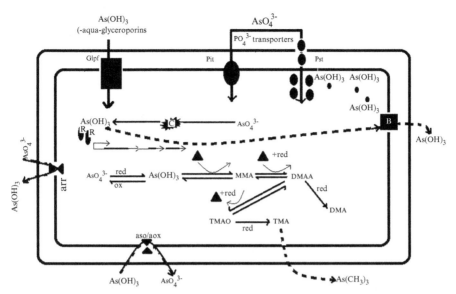

Fig. 2 Microbial processes that are involved in arsenic's detoxification and transport (redrawn from Paez-Espino et al. 2009). In *E. coli*, a gram-negative bacterium As(V) enters the periplasmic space via outer membrane porin (PhoE protein) and is transported into the cytoplasm by the Pit or Pst systems. Inside the cell, As(V) is reduced to As(III) by the enzyme arsenate reductase (ArsC). As(V) cannot pass through ArsAB pump; rather, As(III) is pumped out of the cell by the ArsAB efflux ATPase. The *arsRDABC* operon is regulated by the ArsR repressor protein (a transcriptional regulator) and ArsD co-regulator protein. In gram-positive bacteria, ArsA and ArsD are lacking

methylarsines (MMA) in a pure culture of a methanogen, *Methanobacterium bryantii*. The process was later on reported in *Methanobacterium formicicum* (a methanogen), *Clostridium collagenovorans* (a fermentative bacterium), *Desulfovibrio vulgaris*, and *Desulfovibrio gigas* (sulfate-reducing bacteria) (Michalke et al. 2000). MMA is further converted to MMAA (monomethyl arsenic acid), DMAA (dimethyl arsinic acid), and TMAO (trimethylarsine oxide) (Woolson 1977; Cullen and Reimer 1989; Gadd 1993; Turpeinen 2002). The demethylating microbes reconvert methylated species back to inorganic forms (Sohrin et al. 1997).

Methylation and demethylation play a significant role in the toxicity and mobility of As in soils and groundwater (Wang and Mulligan 2006). As(III) and As(V) methylation may lead to the formation of volatile species, which escape from water and soil surfaces into the atmosphere. Accordingly, As(III) is volatilized to arsine (AsH_3), MMAA to monomethylarsine [MMA, $AsH_2(CH_3)$], DMAA to dimethylarsine [DMA, $AsH(CH_3)_2$], and TMAO to trimethylarsine [TMA, $As(CH_3)_3$] (Cullen and Reimer 1989). Microbial activities also result in volatilization of water-soluble arsenic species in the form of gaseous arsines (AsH_3) (Bachofen et al. 1995; Gao and Burau 1997) (Fig. 1). Methylated species are major source of atmospheric arsines (Gao and Burau 1997). However, the "lifetime" of arsines in air is short, and they were rapidly converted back to water-soluble species, As(V), and trimethyl arsine oxide (TMAO) (Turpeinen 2002).

The reduction of As(V) to As(III) and subsequent methylation of As(III), with trimethylarsine (TMA) as a final product lead to the loss of arsenic from soil (Woolson 1977). In the atmosphere, due to oxidation, volatilized species are reconverted into inorganic species, which are redeposited to soil by rain or by dry deposition (Pongratz et al. 1995).

However, contrary to reports that methylated As species are less toxic than inorganic forms, methylated As(III) species were found to be more damaging to DNA, in the order DMAA(III)>MMAA(III)>[As(III), As(V)]>MMAA(V)> DMAA(V)>trimethylarsine oxide [TMAO(V)] (Ahmad et al. 2002).

3.4 Mobilization and Immobilization

Arsenic present within the subsurface aquifers is mainly natural in origin (not as anthropogenic pollutants). Microbes facilitate the mobilization of arsenic into the aqueous phase (Oremland and Stolz 2005). The physicochemical conditions that favor arsenic mobilization in aquifers are variable, complex, and poorly understood. Siderophores produced by microbes help to mobilize arsenic through the formation of As-siderophore complex (Zawadzka et al. 2007). Mobilization occurs under strongly reducing conditions, in which arsenic is present mainly as As(III), and is released by desorption from or dissolution of iron oxides. Under oxidizing conditions, high concentrations of arsenic have been reported from areas having high groundwater pH (>8) (Smedley and Kinniburgh 2002). In such environments that are dominated by As(V) species, arsenic concentration is positively correlated with anion-forming species (e.g., HCO_3^-, F^-, H_3BO_3, and $H_2VO_4^-$). Immobilization of arsenicals is also possible under reducing conditions. For example, sulfate-reducing microbes that can respire As(V) lead to the formation of As_2S_3 precipitate. Immobilization of arsenic also occurs during the formation of iron sulfides (Plant et al. 2004).

3.5 Microbial Arsenic Uptake and Extrusion

The oxyanions of arsenic enter bacterial cells via transporters involved in the transport of other molecules (Stolz et al. 2006). In general, microbes possess two types of metal uptake systems. The first one is a fast unspecific transport system driven by the chemo-osmotic gradient across the membrane. The mechanism is operative for a variety of metals and is constitutively expressed (Nies 1999). The second system is slow and highly substrate specific. It uses ATP and operates only when needed (Nies and Silver 1995). Owing to the activity of the first system (i.e., unspecified

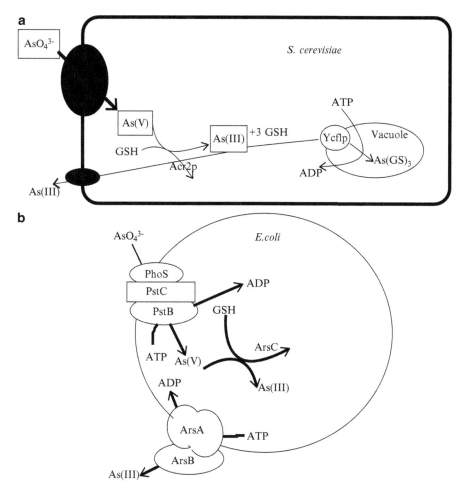

Fig. 3 The mechanism by which arsenic is detoxified in eukaryotic (**a**) and prokaryotic microorganisms (**b**) (Redrawn from Rosen 2002)

transport), microbes often accumulate toxic concentrations of metal ions within their cells. Structural analogs of a particular metal may fool the organism's uptake system (e.g., arsenate and phosphate).

In both prokaryotes and eukaryotes, arsenate enters into cells via phosphate transporters, while arsenite enters cells via aquaglyceroporins and hexose permeases (e.g., yeast and mammals). In a solution with plenty of arsenate (H_3AsO_4), an analog of phosphate, arsenate is taken up via phosphate transporters (Rensing et al. 1999). For example, in *Escherichia coli*, As(V) enters into cells via ATP-coupled Pst pumps. Once inside the cell, As(V) is reduced to As(III) by the enzyme arsenate reductase. The resulting As(III) is exported through an exclusive system composed of single efflux protein which cannot export As(V) (Fig. 3a, b).

Arsenic toxicity can also be avoided by minimizing the uptake of arsenic into cells. *E. coli* possess two different phosphate transport systems—the Pit and the Pst systems (Fig. 2). The Pit system has low phosphate specificity, and it also transports arsenate. In contrast, the Pst system exhibits high phosphate specificity (Rosenberg et al. 1977). Organisms lacking the Pit system exhibited increased resistance to arsenate (Cervantes et al. 1994; Jackson et al. 2003; Bhat 2007) compared to those utilizing the Pst system. Furthermore, arsenate uptake and resulting toxicity could be reduced by increasing the amount of phosphate in the environment (Abedin et al. 2002; Jackson et al. 2003; Bhat 2007).

At neutral pH, As(III) is dissolved to form As(OH)$_3$ acid. However, at alkaline pH, it is present as anion arsenite. There are three types of transporters reported in microbes for the uptake As(III) (Fig. 3a, b). In *E. coli*, a gram-negative prokaryote, the transporter is GlpF (glycerol facilitator), which is a member of aquaporin (AQP) superfamily. However, in eukaryotes, the transporters are Fps1p (the yeast homologue of GlpF) and LmAOP1 (from the parasitic *Leishmania major*). The *E. coli* GlpF was the first transporter identified for As(III) and Sb(III) uptake. It was also the first member of the aquaporin (AQP) superfamily to be identified, even before the human water channel Aqp1 transporter was identified (Liu et al. 2004). This superfamily has two branches: the classical aquaporin, which are water channels with small openings, and aquaglyceroporins that have sufficient channel width for molecules as large as glycerol (Liu et al. 2002). AqpS is a membrane protein belonging to major intrinsic protein or aquaporin superfamily (King et al. 2004). It shows sequence homology with the bacterial glycerol facilitator (GlpF), yeast aquaglyceroporin Fps1, and mammalian aquaglyceroporin AQP9.

Fps1p, a homologue of GlpF, helps *Saccharomyces cerevisiae* in arsenite uptake. Another group that facilitate arsenite uptake is the Hxt glucose transporter (a permease), with uptake rate comparatively slow to that of Fps1p. The presence of glucose competitively inhibits arsenite uptake by Hxts; in such situations, the bulk of arsenite is taken up by the Fps1p (Rensing and Rosen 2009). But, in the absence of glucose in media, nearly 75% of the arsenite enters into yeast through Hxt, only 25% via Fps1p. Other sugar transporters may also be involved in arsenite uptake. In a GlpF⁻ mutant of *E. coli*, arsenite uptake was reported to be about 20% of the uptake of wild type, suggesting involvement of one or more unidentified arsenite uptake systems (Liu et al. 2002; Kamiya et al. 2009; Wu et al. 2010).

To reduce arsenite, the reducing potential is provided by glutathione and glutaredoxin, which is finally extruded from the cells by ArsB alone in prokaryotes and Acr3p in eukaryotes (Bhattacharjee and Rosen 2007). To confirm the involvement of AqpS in arsenic toxicity resistance, the *aqpS* and *arsC* genes of *Sinorhizobium meliloti* were intentionally disrupted individually. Disruption of *aqpS* resulted in increased tolerance to arsenite, but not arsenate. However, cells with disrupted *arsC* showed sensitivity to arsenate. In *S. meliloti*, arsenate enters into cells via phosphate transporters and is reduced to arsenite by ArsC and is then extruded out of the cells by downhill movement through AqpS. AqpS and ArsC together form a novel pathway of As(III) detoxification in *S. meliloti* (Yang et al. 2005; Bhattacharjee and Rosen 2007).

4 The Ars Operon and Proteins

4.1 Microbial Arsenic Sensing

Detecting environmental fluxes of metals is crucial for the survival of any organisms. Metal-responsive transcriptional factors help in maintaining homeostasis between cellular metabolism and environmental availability of metals. To date, seven (ArsR-SmtB, MerR, CsoR-RcnR, CopY, DtxR, Fur, NikR) families of soluble cytoplasmic DNA-binding, metal-sensing transcriptional regulators have been reported to exist in bacteria. Responses to different metals are regulated by different members of each family, which discriminate between metals and trigger expression of appropriate genes. Representative of the different families may also be involved in sensing the same metal (Waldron and Robinson 2009; Osman and Cavet 2010). Metal sensors either act upon a single gene target or as global regulators to alter the expression of multiple genes in response to a particular metal. In addition to soluble metal-sensing transcriptional regulators, individual metal-sensors of structurally distinct families of regulators (e.g., TetR family zinc sensor, ModE molybdate-sensor) and multiple two-component histidine kinase-response regulator systems are also involved in sensing of metal fluxes (Osman and Cavet 2010).

Members of the ArsR-SmtB (ArsR, arsenic regulator; SmtB, *Synechococcus* metallothionein regulator) family sense metals other than the arsenicals, to wit, antimony (Sb) and bismuth (Bi) (Xu et al. 1996). The ArsR protein acts as derepressor, wherein arsenic-binding (+3 oxidation state oxyanions of As, Sb and Bi) impairs DNA binding to alleviate repression (Osman and Cavet 2010). Genes encoding ArsR-SmtB sensors are widespread and have been reported in many bacteria (Campbell et al. 2007).

The ArsR-SmtB family contains nine distinct metal-sensing motifs ($\alpha3$, $\alpha3N$, $\alpha5$, $\alpha3N\alpha5$, $\alpha5C$, $\alpha5$-3, $\alpha4C$, $\alpha3N$-2, $\alpha5$-4) and one nonmetal-sensing motif ($\alpha2\alpha5$) for the detection of different metals (e.g., As, Sb, Bi, Zn, Cd, Pb, Co, Ni, Cu, and Ag) (Campbell et al. 2007; Osman and Cavet 2010). Apo-SmtB is a homodimer with winged helix structure. Helices $\alpha3$ and $\alpha4$ form helix-turn-helix DNA-binding regions. Binding of As to this region impairs DNA binding, resulting in reduced stability of ArsR–DNA complex (Arunkumar et al. 2009). Further, two pairs of metal-binding sites ($\alpha3N/\alpha3N$-2 and $\alpha5/\alpha5$-4) are located in each dimer, at dimer interfaces. The $\alpha3N$ sites are not obligatory for metal sensing. The $\alpha3$ helix contains three cysteine legends forming a trigonal metal-binding site. Although, binding to only two is essential for conformational changes, all three cysteines contributing to form a single subunit of homo-dimer (Shi et al. 1996). The $\alpha3$ cysteines are essential for Ars-induced responsiveness (Shi et al. 1994, 1996; Qin et al. 2007). In *E. coli* ArsR, ligands corresponding to the $\alpha5$ metal-sensing site are lacking. The ArsR (also AseR)-mediated arsenic sensing triggers expression of the *ars* arsenical resistance operon (encoding arsenical-translocating ATPase) (Xu et al. 1996).

4.2   Ars Operon and Transcriptional Regulation

Arsenic toxicity resistance genes are organized in operons present on the nuclear and plasmid DNA of bacteria (Chen et al. 1986; Carlin et al. 1995; Rosen 1996; Bruins et al. 2000). The operon consists of either three (*arsRBC*) or five (*arsRD-ABC*) genes that are organized in a single transcriptional unit (Silver and Phung 1996). The *arsA* and *arsB* genes encode the pump protein that exports As(III) (i.e., help in lowering the intracellular concentration of arsenic), *arsC* to the enzyme arsenate reductase, and ars*R* to the regulatory protein controlling the expression of the *ars* operon. In three-gene operon system, *arsR* encodes the arsenic transcriptional repressor, *arsB* to arsenite permease, and *arsC* to the arsenate reductase (Götz et al. 1983; Cai et al. 1998; Diorio et al. 1995). The ars operons participate in the biogeochemical cycle of arsenic in the sense that they regulate its speciation and mobility (Paez-Espino et al. 2009).

In contrast to the plasmid-borne *ars* operon, the chromosomal *ars* operon contains neither the *arsD* nor the *arsA* cistron (Diorio et al. 1995). The ArsR regulatory proteins are the members of ArsR family of arsenic responsive repressors. The first identified member of this family was 117-residue As(III)-responsive ArsR repressor of the *arsRDABC* operon of plasmid R773. Its more than 200 homologues have been identified in bacteria and archaea (Rensing and Rosen 2009). The detailed structure of ArsR repressors has been provided elsewhere (Rensing and Rosen 2009). Disruption of the chromosomal *ars* operon by *MudI* insertion increased the sensitivity of *E. coli* to sodium arsenite and sodium arsenate toxicity by approximately 10- to 100-fold (Silver et al. 1981). Kostal et al. (2004) reported that overexpression of *ArsR* that shows high affinity and selectivity to arsenite increased the accumulation of arsenic. The bacterial ArsR (*Acidithiobacillus ferrooxidans*) has two vicinal cysteine located near the C-terminus of the protein. However, the involvement of cysteine pairs in metalloid binding remains to be determined.

In the *E. coli* plasmid R773, a 117-residue As(III)/Sb(III)-responsive ArsR repressor is encoded by the *ars RDABC*. Each subunit of the R773 ArsR repressor has a three metal-binding domains—Cys-32 for As(III), Cys-34 for Sb(III), and Cys37 (not required for induction and is not present in all homologues) (Shi et al. 1994). ArsR shows S3 binding contributed by these three cysteine residue of a single subunit (Shi et al. 1996). The inducer binding site is located either near the DNA-binding domain or at the dimmer interface. It is formed by placement of pairs or triads of cysteine residue within a single subunit or between subunits. It appears that the As(III)-binding sites of R773 ArsR, At ArsR, and Cg ArsR resulted from three independent and relatively recent evolutionary events that build on the same backbone repressor-binding protein. The vicinal cysteine pair is sufficient for strong binding of As(III). The best inducer of the *ars* operon is phenyl arsine oxide, which accommodates only two protein ligands due to phenyl ring. Ligands such as carboxylates, serine hydroxyls, or histidine imidazole nitrogen form weak As(III) binding and could participate. In As(III)-responsive repressor, a pair of cysteine residue is expected to form a metalloid-binding site. ArsR, CadC, and SmtB are heterodimers that repress transcription by binding to DNA in the absence of inducing metal ion. They dissociate from the DNA when metals bind to them, leading to the expression of metal ion-resistance genes.

4.3   Arsenate Reductases (ArsCs)

The arsenate reductases (ArsC) facilitate arsenate reduction, which is a detoxification process. Both membrane bound and cytoplasm soluble ArsCs have been reported in microbes. The respiratory ArsC is membrane bound and is coupled to the respiratory electron transfer chain (Mukhopadhyay et al. 2002; Rosen 2002). The arsenate reductase ArsC (13–16 kDa) is a soluble cytoplasmic enzyme, requiring either reduced glutathione or thioredoxin for its activity (Gladysheva et al. 1994; Ji et al. 1994).

There are three independent and unrelated families of cytosolic arsenate reductase (encoded by *ars* operon) involved in arsenate's resistance to toxicity. They differ in structure and may have derived from convergent evolution. The first family of arsenate reductases includes the nuclear and plasmid-borne R773 ArsCs of *E. coli*. They use glutaredoxin (Grx) and GSH as reductants (Mukhopadhyay and Rosen 2002). R773 ArsC is related to Spx of *Bacillus subtilis*, a transcriptional repressor that interacts with the C-terminal domain of α-subunit of RNA polymerase (RNAP), and is essential for growth under disulfide stress (Zuber 2004). The second family of ArsCs includes *Staphylococcus aureus* plasmid pI258 ArsC (Ji and Silver 1992) and *B. subtilis* chromosomal ArsC. They use thioredoxin (Trx) as a reductant and are distantly related to mammalian low molecular weight (LMW) protein tyrosine phosphatases (PT-Pases) (Messens et al. 1999; Bannette et al. 2001). ArsCs of the third family (e.g., AcrP2 of yeast and LmAcr2P from the parasitic *L. major*) are mainly present in eukaryotic microbes (Bobrowicz et al. 1997; Mukhopadhyay and Rosen 1998; Zhou et al. 2004; Rensing and Rosen 2009). They show similarity to the catalytic domain of the Cdc25 cell cycle protein tyrosine phosphatase.

The cyanobacterium *Synechocystis* sp. strain PCC 6803 contains another variant of arsenate reductase (Li et al. 2003), which is homologue to pI258 ArsC and shows both phosphate and arsenate reductase activities. Its catalytic activity is a combination of both R773 and pI258 ArsC. Like R773 ArsC, *Synechocystis* ArsC requires both GSH and Grx as reducing equivalents, but not thioredoxin (i.e., required for the activity of pI258 enzymes). The R773 enzyme requires only one cysteine for catalysis, but the *Synechocystis* ArsC has three essential cysteines similar to the pI258 enzyme. *Synechocystis* and *S. aureus* enzymes appear to be the products of two independent evolutionary pathways (Li et al. 2003). In *C. arsenatis*, acetate is used as electron donor during arsenate respiration catalyzed by the enzyme arsenate reductase, ArsB. It is a dimmer of 87-kDa ArsA and 29-kDa ArsB subunits and is located in the periplasm. ArsAB is coupled to the respiratory chain and provides energy for oxidative phosphorylation.

4.4   Arsenic Permeases (ArsBs)

In highly reducing inner environment of cells, arsenite rather than arsenate is the dominant form. In *E. coli*, the *ars* operon of R773 encodes an arsenite extrusion pump, which confers resistance to As(III) and Sb(III). This efflux pump (i.e., ArsAB) has two subunits; the ArsA-ATPase is catalytic subunit of the pump, it hydrolyzes ATP, and it drives active transport of As(III)/Sb(III) against their chemical gradients (Kuroda et al. 1997). Although ArsB functions as a membrane anchor for ArsA, it also contains the translocation pathway. It alone can facilitate the extrusion of oxyanion across the membrane by acting as a secondary transporter (Dey and Rosen 1995).

In bacteria, two different families of arsenite permeases have been discovered viz., ArsB and Acr3. ArsB is widespread in bacteria and archaea, consisting of 12 membrane spanning segments (Wu et al. 1992). It is encoded by the R773 *arsRD-ABC* operons (Bhattacharjee and Rosen 2007) and mediates transport of As(III) as an anti-porter that facilitates the exchange of trivalent metalloid for protons. ArsB couples arsenite efflux to the electrochemical proton gradient (Meng et al. 2004). The true substrate of ArsB is a polymer of As(III) or Sb(III) or a copolymer of As(III) and Sb(III). The transporter exhibits lower affinity for As(III) polymers than that of Sb(III) polymers. Moreover, As(III) is inhibitory to the Sb(III) transport, while Sb(III) stimulates As(III) transport. The exact nature of the polymer is not known; however, it could be six membered oxo-bridged rings (Meng et al. 2004).

ArsA is an ATPase that helps in As(III) efflux; necessary energy for the process is provided by the ATP hydrolysis (Tisa and Rosen 1990). ArsB becomes associated with the ArsA-ATPase to form a pump that confers a higher level of arsenite resistance than ArsB alone (Rensing and Rosen 2009). Before efflux, the relatively less toxic As(V) is converted to the more toxic As(III). Therefore, it is possible that the As(III) efflux system was first evolved under reducing environments and subsequently coupled with As(V) reduction to accommodate As(V) toxicity, as the earth atmosphere became oxidative (Rosen 2002). Coexpression of *arsA* and *arsB* genes resulted in more efficient As(III) extrusion than expression of *arsB* alone (Dey and Rosen 1995). Cells expressing only *arsB* genes were able to reduce the intracellular level of As(III) compared to the strains without Ars proteins, reflecting the ability of ArsB to catalyze $As(OH)_3/H^+$ exchange (Meng et al. 2004). Homologues of the bacterial ArsA-ATPase are widespread and have been reported from the members of all three domains—bacteria, archaea, and eukaryotes (Bhattacharjee et al. 2001). MRP1 and Ycf1p are other ATPases that pump glutathione-*S*-conjugates out of the cytosol. MRP1 transports arsenic as a triglutathione conjugate (Leslie et al. 2004), while Ycf1 catalyzes the ATP-driven uptake of As(III)-glutathione conjugate into yeast vacuoles (Ghosh et al. 1999).

Acr3p is another family of arsenite transport protein showing functional similarity to the ArsB but differs in sequences. It is more specific and transports arsenite only (Cai et al. 2009). However, in *Synechocystis*, Acr3p was able to transport both arsenite and antimonite (Lopez-Maury et al. 2003).

4.5 Posttransductional Regulation: ArsA and ArsD

The ArsA, which remains normally bound to ArsB, is a soluble protein in the cytosol (Rosen et al. 1988). The soluble ArsA exhibits ATPase activity stimulated by As(III) and Sb(III). The 583-amino-acid ArsA has two homologues halves, A1 and A2, connected by a short linker. Each half has a consensus nucleotide-binding domain (NBD) at the A1–A2 interface (Zhou et al. 2000). A metalloid-binding domain (MBD) that binds three As(III) or Sb(III) is located about $20A^0$ from the NBDs (Zhou et al. 2000). Each metalloid is bound by two ArsA residues. The binding site is composed of Cys-113 and Cys-422 (Ruan et al. 2006), and a third cysteine (Cys-172) participates in high-affinity binding. Binding of As(III) or Sb(III) brings two halves of ArsA together and activates ATP hydrolysis. ArsA has two signature sequences serving as signal transduction domains (STDs), viz., D^{142} TAPTGH^{148}TIRLL in A1 (STD1) and D^{147}TAPTGH^{453}TLLL in A2 (STD2). They correspond to switch II region of many other nucleotide-binding proteins and are supposed to be involved in transmission of energy generated by the hydrolysis of ATP to metalloid transport (Zhou and Rosen 1997). Asp142 and Asp447 are two Mg^{2+} ligands in NBD1 and NBD2, respectively. Whereas, His148 and His453 are Sb(III) ligands in the MBD (Zhou et al. 2000). The As(III)-binding site is located in the first ($\alpha3$) helix of the helix-loop-helix DNA-binding domain. The sulfur atoms of three cysteines are linearly arranged along the $\alpha3$ helix, more than 10 A^0 from Cys32 to Cys37. To bring the cysteine thiolates close to each other, binding of As(III) induces a conformational change of high magnitude that could break the helix, resulting in dissociation of ArsR from the operator/promoter site and transcription of the resistance genes (Zhou et al. 2000).

The product of the *arsD* gene is a 120-residue (a functional homo-dimer) chaperone, which transfers the trivalent metalloids As(III) and Sb(III) directly to the ArsA subunit of the ArsAB pump (i.e., ArsAB As(III)-translocating ATPase) (Rosenzweig 2002; Rensing and Rosen 2009). It is a type of metallochaperone (i.e., arsenic chaperone). Metallochaperones are proteins reported from all three kingdoms and buffer cytosolic metals and deliver them to protein targets such as metalloenzymes and extrusion pumps. In each *ars* operon, the *arsD* gene precedes *arsA* gene, indicating their coevolution for common and related function.

Coexpression of *arsD* gene with *arsAB* resulted in reducing As(III) accumulation in the cell (Lin et al. 2006). The competitive advantage provided by the interaction of ArsD with ArsA was a driving force for the coevolution of these two genes. ArsD has higher affinity for the metalloid than arsA. It protects cells from free metalloid, helps to deliver metalloid to ArsA, and enhances the ATPase activity of ArsA at low metalloid concentrations. There are more than 50 bacterial and archaeal arsenic resistance operons and gene clusters with *arsA* and *arsD* genes, which are always found together in *ars* operons. The order of the genes may differ, but *arsD* gene always precedes *arsA* gene. This linkage suggests that *arsD* and *arsA* genes coevolved before their association with *arsB* genes. And the *arsDA* gene moved laterally into *ars* operon as a unit. The *arsD* gene has a biochemical function related to *ArsA* in arsenic detoxification.

In a mixture of *ArsD* and *ArsA* treated with dibromobimane (dBBr, a fluorogenic homo-bifunctional thiol-specific cross-linking reagent), a cross-linked species (~90 kDa) was detected. The predicted mass of an ArsD-ArsA cross-linked product was increased by addition of MgATP. These results suggest that arsD and arsA interact at their cysteine-rich metalloid-binding sites, and arsD interacts with nucleotide bound form of ArsA (Kosower et al. 1980).

In a mixed culture of cells expressing either *arsDAB* or *arsAB* grown in the presence of sub-toxic concentration of As(III), the cells expressing all three genes removed As(III) from the culture within a week. This showed that cells with *arsDAB* have increased fitness in low As(III) compared to the cells only with *arsAB*. Thus, expression of *ArsD* increases the ability of the arsAB pump to extrude As(III). ArsD and ArsA are chemically cross-linked in a 1:1 complex through the cysteine residue of their metalloid-binding sites. The rate of dissociation of metalloid from ArsD is enhanced four times by its interaction with ArsA, transferring As(III) from the chaperone to the ATPase. ArsD increases the affinity of ArsA for As(III) without altering the V*max* value. This makes ArsAB pump more effective at low but environmentally relevant concentrations of metalloid (10 µM Arsenite) (Lin et al. 2006). Though the molecular mechanism of chaperone activity is not known, it is proposed that ArsD and ArsB bind to the same site on ArsA, and metal is transferred sequentially from ArsD to ArsA to ArsB, with concomitant ATP binding and hydrolysis (Ajees et al. 2011).

ArsR and ArsD are regulatory components acting as a transcription repressor and regulator of the upper limit for operon activity, respectively (Rosen 2002). The regulatory proteins have extremely high affinity for As(III), which bind to their cysteine residues, resulting in an altered DNA binding for transcriptional activation (Rosen 1999).

The ArsD (e.g., R773) has three vicinal cysteine pairs — Cys12–Cys13, Cys112–Cys113, and Cys119–Cys120. Each cysteine pair forms an independent metalloid-binding site (Sun et al. 2001). Cys12–Cys13 and Cys18 form a three-coordinate thiolate metalloid-binding site termed as "MBS1," Cys112–Cys113 (MBS2), and Cys119–Cys120 (MBS3), respectively (Lin et al. 2007; Yang et al. 2010). Only MBS1 participates in metalloid transfer to ArsA and in the activation of ArsA-ATPase activity. The fact the MBSs in ArsA and MBS1 in ArsD are involved in interaction suggests that metalloid-binding sites in both proteins have to be in close proximity, allowing transfer of metalloid directly from the binding site on ArsD to the binding site an arsA. The detail of the metalloid transfer mechanism is not yet known.

4.6 ArsM and ArsH

In both archaea and bacteria, there are gene homologues to the human arsenic methylase. A 283-residue *ArsM* (Arsenite *S*-adenosyl methyltransferase, 29.6 kDa) gene, which helps in methylation, was isolated from the soil bacterium *Rhodopseudomonas palustris* and was cloned in an arsenic-hypersensitive strain of *E. coli*, resulting in conversion of As(III) in DMA(V), TMAO(V), and TMA(III) (Qin et al. 2006). Results showed that the transformation conferred As(III) resistance to the transformed *E. coli*, which had no other *ars* genes, indicating that methylation is sufficient to detoxify arsenic (Qin et al. 2006). TMA(III) is more toxic than arsenite; however, due to its gaseous nature, it volatizes after formation. Hence, there is no accumulation in either cells or media (Qin et al. 2006).

ArsH is another class of arsenate reductase. It has conserved domains related to the NADPH-dependent flavin mononucleotide reductase class of proteins. Though disputed, it confers resistance to arsenicals, but the mechanism is not clear (Rensing and Rosen 2009).

5 Bioremediation of Arsenic-Contaminated Environments

Arsenic toxicity of soil is mainly a function of the arsenic species present and their bioavailability. As(V) is the main species often detected in soils (Cullen and Reimer 1989). Hydrated ions are more toxic than are As complexes/species that are associated with colloidal particles (Russeva 1995; La Force et al. 2000). Soil properties (e.g., pH, texture, temperature, organic matter content) and conditions (e.g., water logging, redox status, soil and site hydrology, and biotic influences) affect the sorption capacity and arsenic toxicity of soils (Turpeinen 2002). In general, iron oxides/hydroxides are mainly involved in the adsorption of arsenic in both acidic and alkaline soils (Polemio et al. 1982). In oxygenated acidic soils, Fe-arsenate [$Fe_3(AsO_4)_2$] controls arsenic solubility, whereas in anoxic soils, it is sulfide of As(III) that controls the concentration of the dissolved arsenic in soil solution. In addition, direct precipitation of arsenic as discrete arsenic solid phases such as reduced arsenopyrite or oxidized hematite also occurs in contaminated soils (Turpeinen 2002). These processes indicate that arsenic becomes progressively less soluble and unavailable with time (Alexander 2000).

It is very important to have efficient and environment-friendly technologies to remove arsenic from contaminated resources. Current techniques used to remove arsenic from contaminated waters (e.g., reverse osmosis, filtration, adsorption, ion exchange, chemical precipitation, distillation) are effective but are costly and time-consuming. Often are unaffordable in parts of the world where As contamination is most rampant (Mohapatra et al. 2007, 2008). In addition, such processes are often hazardous to workers and capable of producing secondary wastes (Lombi et al. 2000).

Bioremediation is method for removing metal residues, including As, relying on the use of plants, microorganisms, and their enzymes to reclaim contaminated natural environments. The approach holds great potential, as it is more efficient than chemical and physical methods (Lloyd et al. 2003).

Arsenic-resistant and/or arsenic-accumulating microbes are widespread among different environments (Merrifield et al. 2004; Takeuchi et al. 2007). Corsini et al. (2010) reported 19 arsenic-resistant bacteria, mostly belonging to the facultative aerobic genera *Bacillus*, *Paenibacillus*, *Staphylococcus*, and to *Rhodococcus* and *Micromonospora*. Biotransformation is an integral part of the biogeochemical cycles of metals and is employed by microbes to get rid of toxic heavy metals. Besides, microbial transformations reduce the toxicity and bioavailability of arsenic not only to themselves but also to other organisms. Therefore, bioremediation could be used as a biological tool to remediate arsenic-contaminated environments. Both prokaryotes and eukaryotic microbes use similar strategies to deal with arsenic poisoning. These include redox conversion, compartmentalization, active efflux pumps, extrusion and immobilization (i.e., intracellular chelation by various metal-binding peptides including glutathione (GSH) phytochelatin and metallothioneins), and reducing the sensitivity of cellular targets (Ji and Silver 1995; Nies and Silver 1995; Nies 1999; Bruins et al. 2000; Turpeinen 2002). The strategies are used either individually or in combination. Microbially transformed, less mobile As species could be sorbed on a mixed solid phase. However, such immobilization is reversible and sorbed metal could be freed by physicochemical factors over time.

In nature, microbes such as methanogens, fermentative bacteria, and sulfate- and iron-reducing bacteria facilitate the anaerobic reduction of arsenate (Christensen et al. 2001). Residual organic matter present in sediments, organic contaminants (e.g., benzene, toluene, ethylbenzene, xylene), and organic acids serves as electron donors (Wang and Mulligan 2006). Species of *Geobacter*, a dominate microbial form involved in Fe(III) reduction in freshwater aquifers, have been identified in West Bengal (Islam et al. 2004), Cambodian (Rowland et al. 2007), and American river floodplain sediments (Saunders et al. 2005), exhibit high rates of As(III) release and Fe(III) reduction. As(V) reduction could be facilitated by incubating As(V) coprecipitated with Al hydroxide with *S. barnesii*, under anaerobic condition (Zobrist et al. 2000).

In addition to previously described redox conversion and biomethylation approaches, others could be used to remediate arsenic-contaminated soils, such as biofilm, biosorption, and siderophore-based remediation rhizoremediation (microbially enhances phytoremediation) etc., Iron oxide adsorption is a commonly used process to remove arsenic from drinking water. In the process, arsenic species are adsorbed onto iron oxides resulting in large particles that are filtered out of the water stream. In its biological analog, the treatment process involves attaching microbes to a filter, the microbial deposition of iron oxides resulting in accumulation of arsenic on the filter. Katsoyiannis et al. (2002) and Katsoyiannis and Zouboulis (2004) used fixed-bed up-flow bioreactors to remove arsenic from groundwater by using iron-oxidizing bacteria. Moreover, during the process, As(III) was partially oxidized to As(V) enabling high arsenic removal efficiency (Katsoyiannis and Zouboulis 2004). However, this conversion mediated by ferric ions takes place only in the presence of iron-oxidizing bacteria. According to Harvey and Crundwell (1996), many full-scale projects have been commissioned based on the bacterial leaching of refractory gold ores to remove As. Refractory gold ores contain arsenic mainly in the form of arsenopyrite (FeAsS), and bacterial leaching of ores produces arsenite [As(III)] and arsenate [As(V)] in solution. During bacterial leaching a substantial amount of the arsenic in solution is precipitated as ferric arsenate ($FeAsO_4$) (Harvey and Crundwell 1996).

Adding nitrate (serve as oxidants) to reducing sediment resulted in the formation of Fe^{3+} particles [Fe(III)-(oxy)hydroxide-containing solids], which helps in the sequestration of mobile As(V) and As(III) in sediments (Senn and Hemond 2004). Further, it was reported that the oxidation of iron (not arsenic) supported the sequestration (Gibney and Nüsslein 2007). Iron hydroxides exhibit strong binding affinity for arsenate (Meng et al. 2002). Sulfate-reducing bacteria help in the formation of iron hydroxide plaque (Murase and Kimura 1997). Therefore, addition of sulfur in the rhizospheric region of plants could decrease the availability of As to plants via iron plaque formation (Hu et al. 2007; Purkayastha 2011).

Microbes such as *Herminiimonas arsenicoxydans* produce extracellular polymeric substances, which help in immobilization of metals (Muller et al. 2007). This could be used to develop biofilm based bioremediation reactors (Chang et al. 2006) useful in removing arsenic from contaminated waters (Lieveremont et al. 2009). Arsenic-contaminated soils could be remediated by adding arsenic immobilizing or arsenate-reducing bacterial population into it (i.e., bioaugmentation). For example, addition of methylating fungi (e.g., *Penicillium* sp. and *Ulocladium* sp.) caused volatilization of arsines from the soil (Edvantoro et al. 2004). In addition, if already present in the medium, growth stimulants (nutrients, growth hormones, electron donor, or acceptors) may be added to accelerate the growth of such microbes (i.e., biostimulation). Biosurfactant (e.g., surfactin, sophorolipid with low toxicity and biodegradable) foam technology can be used to deliver nutrients or microbial population into the subsurface (de Koning and Thiesen 2005).

Phytoremediation is a plant-based green technology successfully used to remove a number of metals from contaminated soils (Lombi et al. 2002). There are different processes employed by plants to remediate contaminated environments, notable among these are phytofiltration, hyperaccumulation, phytoextraction, phytoimmobilization, phytostabilization, and phytotransformation.

Phytostabilization is the use of metal-tolerant plants to mechanically stabilize metal-contaminated land to prevent bulk erosion, reduce airborne transport, and leach pollutants (Kramer 2005). In phytoimmobilization, plants decrease the mobility and bioavailability of metals through altering soil factors that lower metal mobility due to the formation of precipitates (i.e., insoluble compounds) and by sorption on roots (Cotter-Howells et al. 1999). Phytovolatilization is the use of plants to volatilize pollutants. Volatilization of arsenic from soil is a natural process (Frankenberger and Arshad 2002). Turpeinen et al. (1999) reported that in the absence of plant roots volatile arsenic account only for small proportions of total arsenic.

Phytoextraction is the use of metal-accumulating plants capable to extract and translocate metals to the harvestable parts (Kramer 2005). Continuous phytoextraction depends on the natural ability of a plant to accumulate, translocate, and tolerate high concentrations of arsenic over the whole growth cycle (Garbisu and Alkorta 2001). Phytoextraction can be accomplished by using either tolerant, high-biomass plant species or hyperaccumulator plant species. Growth and remediation potential has been assessed for cottonwood (*Populus deltoides* Bartr.), cypress (*Taxodium distichum* L.), eucalyptus (*Eucalyptus amplifolia* Naudin, *Eucalyptus camaldulensis* Dehnh., and *Eucalyptus grandis* Hill), and leucaena (*Leucaena leucocephala* L.). All these plants are potential high-biomass species, showing both constitutive and adaptive mechanisms for accumulating or tolerating high arsenic concentration (Gonzaga et al. 2006).

Many plants hyperaccumulate toxic metals within their different parts and could be used for remediation of metal-contaminated sites. Hyperaccumulating plants are those that can take up and concentrate in excess of 0.1% of a given element in their tissue (Brooks 1998). The use of hyperaccumulator plants is advantageous as it result in production of concentrated toxicants, facilitating the final disposal of the contaminant-rich biomass. However, most of the metal-hyperaccumulator plants grow quite slowly and produce low biomass. Whereas, fast growing plants produce high biomass but are usually sensitive to high metal concentrations, except arsenic hyperaccumulator ferns. Thus, the capacity to accumulate and tolerate high metal concentrations in vegetative parts and to produce high amounts of dry matter is not always mutually exclusive (Robinson et al. 1997; Ma et al. 2001).

Among plant group, ferns have been most widely studies for their potential to remove arsenic from contaminated environments (Singh and Ma 2006). Chinese Brake Fern (*Pteris vittata* L.), *Pteris cretica*, *Pteris longifolia*, *Pteris umbrosa*, and *Pityrogramma calomelanos* are some of the examples of As-hyperaccumulating ferns (Han et al. 2001). Though mainly accumulated in fronds, plant species vary in the parts they use to store As (Purkayastha 2011). In phytoremediation, 90% of the arsenic absorbed was in the aboveground biomass, which could be easily removed by frond harvest (i.e., phytoextraction) (Tu et al. 2002). In *P. vittata*, arsenic leaches from fronds as they senesce, returning it to the soil (Tu et al. 2003). This limited the use of *P. vittata* in phytoremediation of arsenic-contaminated soils. Young fern plants were more efficient in removing arsenic than were older fern plants of similar size (Tu et al. 2004). A detailed account of phytoremediation is beyond the scope of this review, which focuses exclusively on the role of microbes in bioremediation of As-contaminated sites.

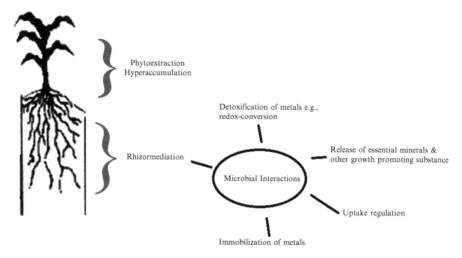

Fig. 4 The different processes that are involved in microbially mediated rhizoremediation of arsenic-contaminated soil

A key to effective phytoremediation, especially phytoextraction, depends on enhanced pollutant availability in the soil solution for plant uptake (Lombi et al. 2002). Rhizospheric factors (e.g., P-fertilizer, rhizospheric microbial population, root exudation, chelating agents) play an important role in improving phytoremediation efficiency (Cao et al. 2003; Khan 2005; Gohre and Paszkowski 2006; Purkayastha 2011). Studies indicate that microbially enhanced phytoextractions could be a promising technology to remediate arsenic-contaminated environments. In rhizosphere, plant-derived exudates may help to stimulate the growth of microbes that subsequently lead to increased degradation of pollutants (Kuiper et al. 2004).

For fern *Pityrogramma calomelanos*, application of P-fertilizer [an analog of As(V)] and rhizosphere bacteria increased biomass as well as As accumulation. However, use of rhizofungi increased biomass but reduced As accumulation (Jankong et al. 2007). Rhizofungi help in phytostabilization of arsenic limiting the As uptake and mobilization (Purkayastha 2011). The rhizofungi-induced protection to As accumulation could occur due to increased availability of phosphate to plants causing increased plant growth and dilution of arsenic or by binding arsenic to fungal mycelia (i.e., rhizospheric immobilization of As) (Sharples et al. 2000; Meharg and Hartley-Whitaker 2002). However, regarding the effect of rhizofungi on As uptake, results are inconsistent even for same plant species (Leung et al. 2006; Trotta et al. 2006).

In Fig. 4, we show the effect of rhizospheric microbial interactions on phytoremediation efficiency of As-contaminated soils and waters. Majority of arsenic-resistant rhizobacteria belongs to group Alphaproteobacteria, Betaproteobacteria, and Gammaproteobacteria and genera such as *Bacillus*, *Achromobacter*, *Brevundimonas*, *Microbacterium*, and *Ochrobactrum* (Cavalca et al. 2010; Purkayastha 2011).

Among rhizospheric microbes, the role of arbuscular mycorrhizal fungi (AMF) in phytoremediation of the As-contaminated sites is well studied (Leyval and Joner 2001). They act as a barrier to the uptake of As by plants (Sharples et al. 2000). However, mycorrhiza-induced phosphate transporters have been reported in different plants, which enhance the As accumulation by infected plants (Liu et al. 2005).

In phytoextraction, aboveground vegetative parts contain a very high amount of toxic metals, which increases the risk of arsenic contamination through food chain. However, in phytostabilization, uptake and mobilization of contaminant is limited and thus is promising and could be applied for long-term remediation of As (Madejon et al. 2002). In these regard, use of rhizospheric AM fungi appears promising. However, the use of rhizospheric AM fungi requires an understanding of plant–soil–microbial interaction, before developing future strategies for phytoremediating As.

In addition to the foraging, use of genetically engineered microbes as selective biosorbents offers an attractive green cure technology for the low cost and efficient removal of arsenic from soil (Singh et al. 2008a, b; Purkayastha 2011). This could be achieved either by expressing metal-binding peptides (Li et al. 2000) or synthetic peptides (Bae et al. 2001). However, their low specificity and affinity for As has restrict their applicability for As remediation (Purkayastha 2011). Most of the effort in this area used microbial gene expressed in plants to enhance arsenic resistance (Dhankher et al. 2002; Song et al. 2003). Dhankher et al. (2002) expressed bacterial ArsC (arsenate reductase) and GSH synthetase (g-ECS) in *Arabidopsis thaliana* resulted in accumulation of As(V) as GSH–As complex. *E. coli* containing phytochelatin synthase (AtPCS) gene from *A. thaliana* and *Schizosaccharomyces pombe* produced PC and accumulated moderate to high levels of arsenic, respectively (Sauge-Merle et al. 2003; Singh et al. 2008b). In yeast, Ycf1p (yeast cadmium factor) plays crucial role in arsenite detoxification. It helps in vacuolar sequestration of arsenite as glutathione conjugates (Rensing and Rosen 2009). Coexpression of PC with glutamylcysteine synthetase (GshI) and arsenic transporter (GlpF) resulted in high PC level and As accumulation (Singh et al. 2008a).

Tsai et al. (2009) coexpressed AtPCS and cysteine desulfhydrase (an aminotransferase-converting cysteine into H_2S) genes in yeast, under anaerobic condition, which led to increased As accumulation due to PC-metal sulfide complex formation. However, requirement of zinc is a major constraint that adversely affects the use of PC for producing cells as a biosorbent (Purkayastha 2011). Overexpression of *ArsR* in *E. coli* showed specific and fast removal of arsenic (i.e., 50 ppb As(III) within 1 h) (Kostal et al. 2004). Similarly, overexpression of a novel As(V)-resistance gene (*arsN*) isolated from industrial treatment plant sludge, showed high arsenic resistance in *E. coli* (Chauhan et al. 2009). Use of resting cells as biosorbent is another low-cost and effective mean for As removal (Singh et al. 2008b).

Microbe-mediated remediation of As-contaminated sites is more or less conceptual. Most of the above studies are performed in laboratory and are yet to be applied in field conditions. Furthermore, we do not know the effectiveness of microbial bioremediation for As or the degree to which the different forms of As are accumulated by microbes. Many arsenic species present a serious challenge to transformation by microbes. For example, insoluble sulfides and uncharged arsenite [$As(OH)_3$] that are predominate under reducing environmental conditions are difficult to remove (Smedley and Kinniburgh 2002). Moreover, we need to better understand the effect of other elements on arsenic's microbial transformation, since the transformation of As is inhibited by the presence of phosphate (Huysmans and Frankenberger 1991) and antimony (Andrewes et al. 2000) and is enhanced by the presence of molybdate (Oremland et al. 2000). Nonetheless, an understanding of biochemical, molecular and genetic mechanisms of resistant microbes could guide us to develop eco-friendly and cost-effective remediation techniques.

6 Arsenic Biosensors and Measurement of Arsenic Bioavailability

Owing to their toxicity and prevalence in drinking water, measurement of inorganic arsenic species in soil and water has received considerable attention over the past decade. Information on the bioavailable fraction is important while assessing the true risk imposed by arsenic. Mays and Hussam (2009) have reviewed the various volumetric methods used to measure inorganic arsenic in water samples.

Biosensors are biological-based devices (whole organism, enzyme-substrate, transcription factor-promoter, etc.) that can be used to quantitatively detect (dose-dependent detectable response) chemicals in the environment. They provide a real-time and specific measurement of the biologically available (hence, ecologically relevant) concentrations of the contaminants, rather than total concentration measured by physicochemical methods, without the need of sophisticated instruments. Microbes are capable of sensing different species of arsenic and their concentrations and have been widely employed to measure environmental concentrations of arsenic species (Sticher et al. 1997; Taurianen et al. 1999; Leveau and Lindow 2002; Belkin 2003; Harms et al. 2006).

Presently, there are two possible designs of biosensors. In the first design known as "light on," the substance to be detected inhibits some aspect of cell metabolism, resulting in a quantifiable response that can be easily detected (Belkin 2003). Another design known as "light off" is based on the use of reporter gene (e.g., *luc* gene or GFP) that is under the control of a constitutive promoter (e.g., *tac* promoter). Such sensors produce a constitutive signal which decreases with the presence of target substance (e.g., arsenic) (Belkin 2003).

A successful biosensor shows strong affinity to its metal effector and weak affinity for other metals. Regulated resistance mechanisms exhibited by microbes against various heavy metals and metalloids have been used to construct whole-cell living biosensors or bioreporters. Biomarkers such as fluorescent and luminescent marker and reporter genes have been used to construct biosensors. A reporter gene encodes for a protein that produces a detectable cellular response. Hence, determines the sensitivity and detection limits of the biosensor (Daunert et al. 2000). The *luc* operon from the firefly *Photinus pyralis* is the most commonly used reporter gene. The degree of expression of reporter gene (sensing element) determines the detectable concentrations (sensitivity) and specificity (detected metals) of the biosensor.

Microbial ability to sense toxic substances (i.e., contaminant-sensing components) is combined with reporter genes to construct biosensors. Although the binding capacity (metal species and sensed concentrations) of the transcription factor used to sense an element influences both sensitivity and specificity, these factors also depend greatly on the status of microbial homeostasis/resistance systems. Understanding the arsenic-resistant mechanism is essential for developing an appropriate biosensor and its response (Strosnider 2003).

All arsenic biosensors are triggered by arsenic (the analyte), which upon entry activates the transcription of the resistance gene, and are followed by the transcription of a reporter gene (Strosnider 2003). The entire resistant gene is not required. Therefore, many biosensors use only the beginning components such as the promoter. The promoters and regulatory genes of the *ars* operons have been used to construct the arsenic specific biosensors. Tauriainen et al. (1999) constructed an arsenic detecting biosensor by expressing firefly luciferase (*lucFF*) reporter gene to the regulatory unit (*ars* promoter + *ArsR* gene from *E. coli* R773 plasmid) in recombinant plasmid pTOO31 with *E. coli* MC1061 as the host strain. In earlier experiment, they used *ars* operon from *S. aureus* p1258 plasmid to construct the biosensor (Tauriainen et al. 1997). However, recombinant plasmid p1258 was able to detect both As(III) and As(V) with equal strength (Tauriainen et al. 1997). In the absence of arsenic, expression of *lucFF* was repressed, while transcription of the promoter was induced by its presence. Further, the luciferase activity was directly related to the concentration of arsenic in the environment. The *E. coli* MC1061 with pTOO31 was able to detect 0.1 μm concentration of As(III) and 0.5 μm of As(V) (Turpeinen et al. 2003). Bacterial luciferase (*luxAB*) reporter gene was used by Corbisier et al. (1993) and Cai and DuBow (1997). Petänen et al. (2001) found *Pseudomonas fluorescens* as a better host compared to that of *E. coli* for determining the bioavailability of arsenic in contaminated soils. Roberto et al. (2002) developed an *E. coli*-based biosensor by coupling *arsR*, *arsD*, and their promoter with reporter gene encoding green fluorescent protein (GFP) from the marine jellyfish, *Aequorea victoria*. The biosensor was able to detect both As(III) and As(V) in a range of 1–10,000 ppb.

Microbial biosensors have been used to measure the bioavailability of essential elements such as nitrate (Mbeunkui et al. 2002), phosphate (Gillor et al. 2002), and iron (Porta et al. 2003). Cai and Dubow (1997) applied a bacterial sensor to estimate the toxicity of an arsenic-containing wood preservative (i.e., chromate copper arsenate). However, use of microbial biosensors to assess the bioavailability of arsenic in soil samples is complicated, since the response of biosensors may be biased by the presence of soil particles. Though these particles can be filtered out from samples, this may produce an underestimation of arsenic bioavailability because certain arsenic fractions (e.g., arsenic complexes with colloidal and fine particles) are eliminated during filtration.

Although biosensors have many advantages, they also suffer from many disadvantages (short lifetime, lack of genetic stability, etc.) (Strosnider 2003). Additional research is required to improve the design and sensitivity of biosensors that may be useful in detecting and measuring arsenic.

7   Conclusions

After performing this comprehensive literature review, we have drawn certain conclusions, as follows:

1. Microbes are important constituents of the global arsenic cycle and play a key role in regulating the speciation and cycling of different forms of arsenic in the environment.
2. They transform highly toxic arsenic species into less toxic forms, which reduces arsenic toxicity not only to the microbes themselves but to other species as well.
3. Microbe-based biosensors offer improvements for detecting the presence of arsenicals in contaminated soil and water, as well as their bioavailability.
4. An understanding of the mechanism(s), proteins and genes employed by microbes to address arsenic toxicity will help in developing arsenic-resistant agriculturally important plants and microbes.
5. Nonetheless, despite many trials, microbes have not been used to bioremediate arsenic under actual natural field conditions.

8   Summary

Arsenic (As) is a nonessential element that is often present in plants and in other organisms. However, it is one of the most hazardous of toxic elements globally. In many parts of the world, arsenic contamination in groundwater is a serious and continuing threat to human health. Microbes play an important role in regulating the environmental fate of arsenic. Different microbial processes influence the biogeochemical cycling of arsenic in ways that affect the accumulation of different arsenic species in various ecosystem compartments. For example, in soil, there are bacteria that methylate arsenite to trimethylarsine gas, thereby releasing arsenic to the atmosphere. In marine ecosystems, microbes exist that can convert inorganic arsenicals to organic arsenicals (e.g., di- and tri-methylated arsenic derivatives, arsenocholine, arsenobetaine, arsenosugars, arsenolipids). The organoarsenicals are further metabolized to complete the arsenic cycle.

Microbes have developed various strategies that enable them to tolerate arsenic and to survive in arsenic-rich environments. Such strategies include As exclusion from cells by establishing permeability barrier, intra- and extracellular sequestration, active efflux pumps, enzymatic reduction, and reduction in the sensitivity of cellular targets. These strategies are used either singly or in combination. In bacteria, the genes for arsenic resistance/detoxification are encoded by the arsenic resistance operons (*ars* operon).

In this review, we have addressed and emphasized the impact of different microbial processes (e.g., arsenite oxidation, cytoplasmic arsenate reduction, respiratory arsenate reduction, arsenite methylation) on the arsenic cycle. Microbes are the only life forms reported to exist in heavy arsenic-contaminated environments. Therefore, an understanding of the strategies adopted by microbes to cope with arsenic stress is important in managing such arsenic-contaminated sites. Further future insights into the different microbial genes/proteins that are involved in arsenic resistance may also be useful for developing arsenic resistant crop plants.

References

Abedin MJ, Cresser MS, Meharg AA, Feldman J, Cotter-Howells J (2002) Arsenic accumulation and metabolism in rice (*Oryza sativa* L.)[J]. Environ Sci Technol 36:962–968

Abernathy CO, Thomas DJ, Calderon RL (2003) Health effects and risk assessment of arsenic. J Nutr 133:1536S–1538S

Acharyya SK, Chakraborty P, Lahiri S, Raymahashay BC, Guha S, Bhowmik A (1999) Arsenic poisoning in the Ganges delta. Nature 401:545–547

Ahamed S, Sengupta MK, Mukherjee A, Hossain MA, Das B, Nayak B, Pal A, Mukherjee SC, Pati S, Dutta RN, Chatterjee G, Mukherjee A, Srivastava R, Chakraborti D (2006) Arsenic groundwater contamination and its health effects in the state of Uttar Pradesh (UP) in upper and middle Ganga Plain, India: a severe danger. Sci Total Environ 370:310–322

Ahmad S, Kitchin KT, Cullen WR (2002) Plasmid DNA damage caused by methylated arsenicals, ascorbic acid and human liver ferritin. Toxicol Lett 133:47–57

Ajees AA, Yang J, Rosen BP (2011) The ArsD As(III) metallochaperone. Biometals 24:391–399

Alexander M (2000) Ageing, bioavailability and overestimation of risk form environmental pollutants. Environ Sci Technol 34:4259–4265

Amini M, Abbaspour KC, Berg M, Winkel L, Hug SJ, Hoehn E, Yang H, Johnson CA (2008) Statistical modeling of global geogenic arsenic contamination in groundwater. Environ Sci Technol 42:3669–3675

Anderson CR, Cook GM (2004) Isolation and characterization of arsenate-reducing bacteria from arsenic contaminated sites in New Zealand. Curr Microbiol 48:341–347

Andrewes P, Cullen WR, Polishchuk E (2000) Arsenic and antimony biomethylation by *Scopulariopsis brevicaulis*: Interaction of arsenic and antimony compounds. Environ Sci Technol 34:2249–2253

Arunkumar AI, Campanello GC, Giedroc DP (2009) Solution structure of a paradigm ArsR family zinc sensor in the DNA-bound state. Proc Natl Acad Sci U S A 106:18179–18182

Aschengrau A, Zierler S, Cohen A (1989) Quality of community drinking water and the occurrence of spontaneous abortion. Arch Environ Health 44:283–290

Bachofen R, Birch L, Buchs U, Ferloni P, Flynn I, Jud G, Tahedel H, Chasteen TG (1995) Volatilization of arsenic compounds by microorganisms. In: Hinchee RE (ed) Bioremediation of inorganics. Batelle Press, Columbus, OH, pp 103–108

Bae W, Mehra RK, Mulchandani A, Chen W (2001) Genetic engineering of *Escherichia coli* for enhanced uptake and bioaccumulation of mercury. Appl Environ Microbiol 67:5335–5338

Balasoiu C, Zagury G, Deshenes L (2001) Partitioning and speciation of chromium, copper and arsenic in CCA-contaminated soils: influence of soil composition. Sci Total Environ 280:239–255

Bannette MS, Guan Z, Laurberg M, Su XD (2001) *Bacillus subtilis* arsenate reductase is structurally similar to low molecular weight protein tyrosine phosphatases. Proc Natl Acad Sci U S A 98:13577–13582

Beane Freeman LE, Dennis LK, Lynch CF, Thorne PS, Just CL (2004) Toenail arsenic content and cutaneous melanoma in Iowa. Am J Epidemiol 160:679–687

Belkin S (2003) Microbial whole-cell sensing systems of environmental pollutants. Curr Opin Microbiol 6:206–212

Bhat NS (2007) Characterization of arsenic resistant bacteria and novel gene cluster in *Bacillus* sp. CDB3. Ph D thesis, School of biological sciences, University of Wollongong

Bhattacharjee H, Rosen BP (2007) Arsenic metabolism in prokaryotic and eukaryotic microbes. In: Nies DH, Simon S (eds) Molecular microbiology of heavy metals. Springer, Heidelberg, New York, pp 371–406

Bhattacharjee H, Ho YS, Rosen BP (2001) Genomic organization and chromosomal localization of the *Asna1* gene, a mouse homologue of a bacterial arsenic-translocating ATPase gene. Gene 272:291–299

Bhattacharjee H, Mukhopadhyay R, Thiyagarajan S, Rosen BP (2008) Aquaglyceroporin: ancient channel for metalloids. J Biol 7:33

Blair PC, Thompson MB, Bechtold M, Wilson RE, Moorman Fowler BA (1990) Evidence for oxidative damage to red blood cells in mice induced by arsine gas. Toxicology 63:25–34

Bobrowicz P, Wysocki R, Owsianik G, Goffeau A, Ulaszewski S (1997) Isolation of three contiguous genes, *ACR1*, *ACR2*, and *ACR3*, involved in resistance to arsenic compounds in the yeast *Saccharomyces cerevisiae*. Yeast 13:819–828

Botes E, van Heerden E, Litthauer D (2007) Hyper-resistance to arsenic to arsenic in bacteria isolated from an antimony mine in South Africa. S Afr J Sci 103:279–281

Bowell RJ, Parshley J (2001) Arsenic cycling in mining environment. Characterization of waste, chemistry, and treatment and disposal, proceedings and summary report on USA. EPA workshop on managing arsenic risks to the environment, Denver, Colorado, USA, 1–3 May 2001

Brooks RR (1998) Plants that hyperaccumulate heavy metals. University Press, Cambridge, p 380

Bruins MR, Kapil S, Oehme FW (2000) Microbial resistance to metals in the environment. Ecotoxicol Environ Saf 45:198–207

Buchet JP, Lauwerys R (1981) Evaluation of exposure to inorganic arsenic in man. In: Facchetti S (ed) Analytical techniques for heavy metals in biological fluids. Elsevier, Amsterdam, pp 75–89

Cai J, Dubow MS (1997) Use of a luminescent bacterial biosensor for biomonitoring and characterization of arsenic toxicity of chromated copper arsenate (CCA). Biodegradation 8:105–111

Cai J, Salmon K, DuBow MS (1998) A chromosomal *ars* operon homologue of *Pseudomonas aeruginosa* confers increased resistance to arsenic and antimony in *Escherichia coli*. Microbiology 144:2705–2713

Cai L, Liu G, Rensing C, Wang G (2009) Genes involved in arsenic transformation and resistance associated with different levels of arsenic-contaminated soils. BMC Microbiol 9:1–11

Campbell DR, Chapman KE, Waldron KJ, Tottey S, Kendall S, Cavallaro G, Andreini C, Hinds J, Stoker NG, Robinson NJ, Cavet JS (2007) Mycobacterial cells have dual nickel-cobalt sensors: sequence relationships and metal sites of metal-responsive repressors are not congruent. J Biochem (Tokyo) 282:32298–32310

Cao X, Ma LQ, Shiralipour A (2003) Effects of compost and phosphate amendments on arsenic mobility in soils and arsenic uptake by the hyperaccumulator *Pteris vittata* L. Environ Pollut 126:157–167

Carbrey JM, Gorelick-Feldman DA, Kozono D, Praetorius J, Nielsen S, Agre P (2003) Aquaglyceroporin AQP9: solute permeation and metabolic control of expression in liver. Proc Natl Acad Sci U S A 100:2945–2950

Carlin A, Shi W, Dey S, Rosen BP (1995) The *ars* operon of *Escherichia coli* confers arsenical and antimonial resistance. J Bacteriol 177:981–986

Cavalca L, Zanchi R, Corsini A, Colombo M, Romagnoli C, Canzi E, Andreoni V (2010) Arsenic-resistant bacteria associated with roots of the wild *Cirsiumarvense* (L.) plant from an arsenic polluted soil, and screening of potential plant growth-promoting characteristics. Syst Appl Microbiol 33:154–164

Cervantes C, Ji G, Ramirez JL, Silver S (1994) Resistance to arsenic compounds in microorganisms. FEMS Microbiol Rev 15:355–367

Challenger F (1945) Biological methylation. Chem Rev 36:315–361

Chang WC, Hsu GS, Chiang SM, Su MC (2006) Heavy metal removal from aqueous solution by wasted biomass from a combined AS-biofilm process. Bioresour Technol 97:1503–1508

Chang JS, Kim YH, Kim KW (2008) The *ars* genotype characterization of arsenic-resistant bacteria from arsenic-contaminated gold-silver mines in the republic of Korea. Appl Microbiol Biotechnol 80:155–165

Chauhan NS, Ranjan R, Purohit HJ, Kalia VC, Sharma R (2009) Identification of genes conferring arsenic resistance to Escherichia coli from an effluent treatment plant sludge metagenomic library. FEMS Microbiol Ecol 67:130–139

Chen CM, Misra T, Silver S, Rosen BP (1986) Nucleotide sequence of the structural genes for an anion pump: the plasmid-encoded arsenical resistance operon. J Biol Chem 261:15030–15038

Christensen TH, Kjeldsen P, Bjerg PL, Jensen DL, Christensen JB, Baun A, Albrechtsen HJ, Heron C (2001) Biogeochemistry of landfill leachate plumes. Appl Geochem 16:659–718

Clausen CA (2004) Improving the two-step remediation process for CCA-treated Wood: Part II. Evaluating bacterial nutrient sources. Waste Manag 24:407–411

Corbisier P, Ji G, Nuyts G, Mergeay M, Silver S (1993) *luxAB* gene fusions with the arsenic and cadmium resistance operons of *Staphylococcus aureus* plasmid pI258. FEMS Microbiol Lett 110:231–238

Corsini A, Cavalca L, Crippa L, Zaccheo P, Andreoni V (2010) Impact of glucose on microbial community of a soil containing pyrite cinders: role of bacteria in arsenic mobilization under submerged condition. Soil Biol Biochem 42:699–707

Cotter-Howells JD, Champness PE, Charnock JM (1999) Mineralogy of Pb/P grains in the roots of *Agrostis capillaris* L. by ATEM and EXAFS. Mineral Mag 63:777–789

Cullen WR, Reimer KJ (1989) Arsenic speciation in the environment. Chem Rev 89:713–764

Datta DV, Mitra JK, Chhautani PN, Chakravati RN (1979) Chronic oral arsenic intoxication as a possible etiological factor in idiopathic portal hypertension (non-cirrhotic portal fibrosis) in India. Gut 20:378–384

Daunert S, Barrett G, Feliciano JS, Shetty RS, Shresta S, Smith-Spencer W (2000) Genetically engineered whole-cell sensing systems: coupling biological recognition with reporter genes. Chem Rev 100:2705–2738

Dave SR, Gupta KH, Tipre DR (2008) Characterization of arsenic resistant and arsenopyrite oxidizing *Acidithiobacillus ferrooxidans* from Hutti gold leachate and effluents. Bioresour Technol 99:7514–7520

de Koning J, Thiesen S (2005) Aqua Solaris—an optimized small scale desalination system with 40 litres output per square meter based upon solar-thermal distillation. Desalination 182:503–509

DeMel S, Shi J, Martin P, Rosen BP, Edwards BF (2004) Arginine 60 in the ArsC arsenate reductase of *E. coli* plasmid R773 determines the chemical nature of the bound As(III) product. Protein Sci 13:2330–2340

Dey S, Rosen BP (1995) Dual mode of energy coupling by the oxyanion-translocating ArsB protein. J Bacteriol 177:385–389

Dhankher OP, Li Y, Rosen BP, Shi J, Salt D, Senecoff JF, Sashti NA, Meagher RB (2002) Engineering tolerance and hyperaccumulation of arsenic in plants by combining arsenate reductase and gamma-glutamylcysteine synthetase expression. Nat Biotechnol 20:1140–1145

Diorio C, Cai J, Marmor J, Shinder R, DuBow MS (1995) An *Escherichia coli* chromosomal *ars* operon homolog is functional in arsenic detoxification and is conserved in gram-negative bacteria. J Bacteriol 177:2050–2056

Drahota P, Filippi M (2009) Secondary arsenic minerals in the environment: a review. Environ Int 35:1243–1255

Edvantoro BB, Naidu R, Megharaj M, Merrington G, Singleton I (2004) Microbial formation of volatile arsenic in cattle dip site soils contaminated with arsenic and DDT. Appl Soil Ecol 25:207–217

Frankenberger WTJR, Arshad M (2002) Volatilisation of arsenic. In: Frankenberger WT Jr (ed) Environmental chemistry of arsenic. Marcel Dekker, New York, pp 363–380

Gadd GM (1993) Microbial formation and transformation of organometallic and organometalloid compounds. FEMS Microbiol Rev 11:297–316

Gao S, Burau RG (1997) Environmental factors affecting rates of arsenic evolution from and mineralization of arsenicals in soil. J Environ Qual 26:753–763

Garbisu C, Alkorta I (2001) Phytoextraction: a cost-effective plant based technology for the removal of metals from the environment. Bioresour Technol 77:229–236

Gebel T (2000) Confounding variables in the environmental toxicology of arsenic. Toxicology 144:155–162

Ghosh M, Shen J, Rosen BP (1999) Pathways of As(III) detoxification in *Saccharomyces cerevisiae*. Proc Natl Acad Sci U S A 96:5001–5006

Gibney BP, Nüsslein K (2007) Arsenic sequestration by nitrate respiring microbial communities in urban lake sediments. Chemosphere 70:329–336

Gihring TM, Banfield JF (2001) Arsenite oxidation and arsenate respiration by a new *Thermus* isolate. FEMS Microbiol Lett 204:335–340

Gihring TM, Druschel GK, McCleskey RB, Hamers RJ, Banfield JF (2001) Rapid arsenite oxidation by *Thermus aquaticus* and *Thermus thermophilus*: field and laboratory investigations. Environ Sci Technol 35:3857–3862

Gillor O, Hadas O, Post AF, Belkin S (2002) Phosphorus bioavailability monitoring by a bioluminescent cyanobacterial sensor strain. J Phycol 38:107–115

Gladysheva TB, Oden KL, Rosen BP (1994) Properties of the arsenate reductase of plasmid R773. Biochemistry 33:7288–7293

Gohre V, Paszkowski U (2006) Contribution of the arbuscular mycorrhizal symbiosis to heavy metal phytoremediation. Planta 223:1115–1122

Gonzaga MIS, Santos JAG, Ma LQ (2006) Arsenic phytoextraction and hyperaccumulation by fern species. Scientia Agricola 63:90–101

Götz F, Zabielski J, Philipson L, Lindberg M (1983) DNA homology between the arsenate resistance plasmid pSX267 from *Staphylococcus xylosus* and the penicillinase plasmid p1258 from *S. aurefl*. Plasmid 9:126–137

Green HH (1918) Description of a bacterium which oxidizes arsenite to arsenate and of one which reduces arsenate to arsenite, isolated from a cattle-dipping tank. S Afr J Sci 14:465–467

Guha Muzumdar DN (2006) Mission report to ministry of health and ministry of rural development of Cambodia, on detection, confirmation and management of arsenicosis. A WHO consultancy project report

Guo XJ, Fujono Y, Kaneko S, Wu KG, Xia YJ, Yashimura T (2001) Arsenic contamination of groundwater and prevalence of arsenical dermatosis in the Hetao Plain area, Tumer Mongolia, China. Mol Cell Biochem 222:137–140

Han FX, Banin A, Triplett GB (2001) Redistribution of heavy metals in arid-zone soils under wetting-drying cycle soil moisture regime. Soil Sci 166:18e28

Harms H, Wells MC, van der Meer JR (2006) Whole-cell living biosensor: are they ready for environmental applications? Appl Microbiol Biotechnol 7:1–8

Harvey PI, Crundwell FK (1996) The effect of As(III) on the growth of *Thiobacillus ferrooxidans* in an electrolytic cell under controlled redox potential. Miner Eng 9:1059–1068

Hu ZY, Zhu YG, Li M, Zhang LG, Cao ZH, Smith FA (2007) Sulfur (S)-induced enhancement of iron plaque formation in the rhizosphere reduces arsenic accumulation in rice (*Oryza sativa* L.) seedlings. Environ Pollut 147:387–393

Huysmans KD, Frankenberger WT (1991) Evolution of trimethylarsine by a *Penicillium* sp. isolated from agricultural evaporation pond water. Sci Total Environ 105:13–28

Ilyaletdinov AN, Abdrashitova SA (1981) Autotrophic oxidation of arsenic by a culture of *Pseudomonas arsenitoxidans*. Mikrobiologiya 50:197–204

Islam FS, Gault AG, Boothman C, Polya DA, Charnock JM, Chatterjee D, Lloyd JR (2004) Role of metal-reducing bacteria in arsenic release from Bengal delta sediments. Nature 430:68–71

Jackson CR, Jackson EF, Dugas SL, Gamble K, Williams SE (2003) Microbial transformation of arsenite and arsenate in natural environment. Recent Res Develop Microbiol 7:103–118

Jackson CR, Horrison KG, Dugas SL (2005) Enumeration and characterization of culturable arsenate resistant bacteria in large estuary. Syst Appl Microbiol 28:727–734

Jankong P, Visoottiviseth P, Khokiattiwong S (2007) Enhanced phytoremediation of arsenic contaminated land. Chemosphere 68:1906–1912

Ji G, Silver S (1992) Regulation and expression of the arsenic resistance operon from *Staphylococcus aureus* plasmid pI258. J Bacteriol 174:3684–3694

Ji G, Silver S (1995) Bacterial resistance mechanisms for heavy metals of environmental concern. J Ind Microbiol 14:61–75

Ji G, Garber E, Armes L, Chen C-M, Fuchs J, Silver S (1994) Arsenate reductase of *Staphylococcus aureus* plasmid pI258. Biochemistry 33:7294–7299

Kaise T, Fukui S (1992) The chemical form and acute toxicity of arsenic compounds in marine organisms. Appl Organomet Chem 6:155–160

Kaltreider RC, Davis AM, Lariviere JP, Hamilton JW (2001) Arsenic alters the function of the glucocorticoid receptor as a transcription factor. Environ Health Perspect 109:245–251

Kamiya T, Tanaka M, Mitani N, Ma JF, Maeshima M, Fujiwara T (2009) NIP1;1, an aquaporin homolog, determines the arsenite sensitivity of *Arabidopsis thaliana*. J Biol Chem 284: 2114–2120

Katsoyiannis IA, Zouboulis AI (2004) Application of biological processes for the removal of arsenic from groundwaters. Water Res 38:17–26

Katsoyiannis I, Zouboulis A, Althoff H, Bartel H (2002) As(III) removal from groundwaters using fixed-bed upflow bioreactors. Chemosphere 47:325–332

Khan AG (2005) Role of soil microbes in the rhizospheres of plants growing on trace metal contaminated soils in phytoremediation. J Trace Elem Med Biol 18:355–364

King LS, Kozono D, Agre P (2004) From structure to disease: the evolving tale of aquaporin biology. Nat Rev Mol Cell Biol 5:687–698

Kosower NS, Newton GL, Kosower EM, Ranney HM (1980) Bimane fluorescent labels. Characterization of the bimane labeling of human hemoglobin. Biochim Biophys Acta 622:201–209

Kostal J, Yang R, Wu CH, Mulchandani A, Chen W (2004) Enhanced arsenic accumulation in engineered bacterial cells expressing *ArsR*. Appl Environ Microbiol 70:4582–4587

Kotze AA, Tuffin MI, Deane SM, Rawlings DY (2006) Cloning and characterization of the chromosomal arsenic resistance genes from *Acidithiobacillus caldus* and enhanced arsenic resistance on conjugal transfer of *ars* genes located on transposon Tn*AtcArs*. Microbiology 152:3551–3560

Kramer U (2005) Phytoremediation: novel approaches to cleaning up polluted soils. Curr Opin Biotechnol 16:133–141

Kuiper I, Lagendijk EL, Bloemberg GV, Lugtenberg BJJ (2004) Rhizoremediation: a beneficial plant-microbe interaction. Mol Plant Microbe Interact 17:6–15

Kuroda M, Dey S, Sanders OI, Rosen BP (1997) Alternate energy coupling of ArsB, the membrane subunit of the Ars anion-translocating ATPase. J Biol Chem 272:326–331

La Force MJ, Hansel CM, Fendorf S (2000) Arsenic speciation, seasonal transformations, and co-distribution with iron in a mine waste-influenced palustrine emergent wetland. Environ Sci Technol 34:3937–3943

Laverman AM, Switzer Blum J, Schaefer JK, Philips EJP, Lovley DR, Oremland RS (1995) Growth of strain SE-3 with arsenate and other diverse electron acceptors. Appl Environ Microbiol 61:3556–3561

Leonard A (1991) Arsenic. In: Merian E (ed) Metals and their compounds in the environment. VCH, Weinheim, pp 751–772

Leslie EM, Haimeur A, Waalkes MP (2004) Arsenic transport by the human multi-drug resistance protein 1 (MRP1/ABCC1): evidence that a tri-glutathione conjugate is required. J Biol Chem 279:32700–32708

Leung HM, Ye ZH, Wong MH (2006) Interactions of mycorrhizal fungi with *Pteris vittata* (As hyperaccumulator) in As-contaminated soils. Environ Pollut 139:1–8

Leveau JHJ, Lindow SE (2002) Bioreporters in microbial ecology. Curr Opin Microbiol 5:259–265

Leyval C, Joner EJ (2001) Bioavailability of heavy metals in the mycorrhizosphere. In: Gobran GR, Wenzel WW, Lombi E (eds) Trace elements in the rhizosphere. CRC, Boca Raton, FL, pp 165–185

Li Y, Cockburn W, Kilpatrick J, Whitelam GC (2000) Cytoplasmic expression of a soluble synthetic mammalian metallothionein-alpha domain in *Escherichia coli*-enhanced tolerance and accumulation of cadmium. Mol Biotechnol 16:211–219

Li R, Haile JD, Kennelly PJ (2003) An arsenate reductase from *Synechocystis* sp. strain PCC 6803 exhibits a novel combination of catalytic characteristics. J Bacteriol 185:6780–6789

Lieveremont D, Bertin PN, Lett MC (2009) Arsenic in contaminated waters: biogeochemical cycle, microbial metabolism and biotreatment processes. Biochimie 91:1229–1237

Lin YF, Walmsley AR, Rosen BP (2006) An arsenic metallochaperone for an arsenic detoxification pump. Proc Natl Acad Sci U S A 103:15617–15622

Lin YF, Yang J, Rosen BP (2007) ArsD Residues Cys12, Cys13, and Cys18 form an As(III)-binding site required for arsenic metallochaperone activity. J Biol Chem 282:16783–16791

Liu J, Rosen BP (1997) Ligand interactions of the ArsC arsenate reductase. J Biol Chem 272:21084–21089

Liu J, Chen H, Miller DS, Sauvedra JE, Keefer LK, Johnson DR, Klaassen CD, Waalkes MP (2001) Overexpression of glutathione s-transferase II and multi-drug resistance transport proteins is associated with acquired tolerance to inorganic arsenic. Mol Pharmacol 60:302–309

Liu Z, Shen J, Carbrey JM, Mukhopadhyay R, Agre P, Rosen BP (2002) Arsenite transport by mammalian aquaglyceroporins AQP7 and AQP9. Proc Natl Acad Sci U S A 99:6053–6058

Liu Z, Shen J, Carbrey JM, Agre P, Rosen BP (2004) Arsenic trioxide uptake by human and rat aquaglyceroporins. Biochem Biophys Res Commun 316:1178–1185

Liu T, Golden JW, Giedroc DP (2005) A zinc(II)/lead(II)/cadmium(II)-inducible operon from the cyanobacterium *Anabaena* is regulated by AztR, an alpha3N ArsR/SmtB metalloregulator. Biochemistry 44:8673–8683

Lloyd JR, Lovley DR, Macaski LE (2003) Biotechnological application of metal-reducing microorganisms. Adv Appl Microbiol 53:85–128

Lombi E, Wenzel WW, Adriano DC (2000) Arsenic-contaminated soils: II. Remedial action. In: Wise DL, Tarantolo DJ, Inyang HI, Cichon EJ (eds) Remedial of hazardous waste contaminated soils. Marcel Dekker, New York, pp 739–758

Lombi E, Zhao FJ, Fuhrmann M, Ma LQ, McGrath SP (2002) Arsenic distribution and speciation in the fronds of the hyperaccumulator *Pteris vittata*. New Phytol 156:195–203

Lopez-Maury L, Florencio FJ, Reyes JC (2003) Arsenic sensing and resistance system in the cyanobacterium *Synechocystis* sp. strain PCC 6803. J Bacteriol 185:5363–5371

Luong JHT, Majid E, Male KB (2007) Analytical tools for monitoring arsenic in the environment. Open Anal Chem J 1:7–14

Ma LQ, Komar KM, Tu C, Zhang W, Cai Y, Kennelley ED (2001) A fern that hyperaccumulates arsenic. Nature 409:579

Madejon P, Murillo JM, Maranon T, Cabrera F, Lopez R (2002) Bioaccumulation of As, Cd, Cu, Fe and Pb in wild grasses affected by the Aznalcollar mine spill (SW Spain). Sci Total Environ 290:105–120

Marshall G, Ferreccio C, Yuan Y, Bates MN, Steinmaus C, Selvin S, Liaw J, Smith AH (2007) Fifty-year study of lung and bladder cancer mortality in Chile related to arsenic in drinking water. J Natl Cancer Inst 99:920–928

Martin P, DeMel S, Shi J, Gladysheva T, Gatti DL, Rosen BP, Edwards BF (2001) Insights into the structure, salvation, and mechanism of ArsC arsenate reductase, a novel arsenic detoxification enzyme. Structure 9:1071–1081

Masscheleyn PH, Delaune RD, Patrick WH (1991) Effect of redox potential and pH on arsenic speciation and solubility in a contaminated soil. Environ Sci Technol 25:1414–1419

Mays DE, Hussam A (2009) Volumetric methods for determination and speciation of inorganic arsenic in the environment—a review. Anal Chim Acta 646:6–16

Mazumdar DNG (2008) Chronic arsenic toxicity and human health. Indian J Med Res 128:436–447

Mbeunkui F, Richaud C, Etienne AL, Schmid RD, Bachmann TT (2002) Bioavailable nitrate detection in water by an immobilized luminescent cyanobacterial reporter strain. Appl Microbiol Biotechnol 60:306–312

McBride BC, Wolfe RS (1971) Biosynthesis of dimethylasrine by a methanobacterium. Biochemistry 10:4312–4317

Meharg AA, Hartley-Whitaker J (2002) Arsenic uptake and metabolism in arsenic resistant and nonresistant plant species. New Phytol 154:29–43

Meng X, Korfiatis GP, Bang S, Bang KW (2002) Combined effects of anions on arsenic removal by iron hydroxides. Toxicol Lett 133:103–111

Meng YL, Liu Z, Rosen BP (2004) As(III) and Sb(III) uptake by GlpF and efflux by ArsB in *Escherichia coli*. J Biol Chem 279:18334–18341

Merrifield ME, Ngu T, Stillman MJ (2004) Arsenic binding to *Fucus vesiculosus* metallothionein. Biochem Biophys Res Commun 324:127–132

Messens J, Hayburn G, Desmyter A, Laus G, Wyns L (1999) The essential catalytic redox couple in arsenate reductase from *Staphylococcus aureus*. Biochemistry 38:16857–16865

Messens J, Martins JC, van Belle K, Brosens E, Desmyter A, De Gieter M, Wieruszeski JM, Willem R, Wyns L, Zegers I (2002) All intermediates of the arsenate reductase mechanism, including an intramolecular dynamic disulfide cascade. Proc Natl Acad Sci U S A 99:8506–8511

Michalke K, Wickenheiser EB, Mehring M, Hirner AV, Hensel R (2000) Production of volatile derivatives of metalloids by microflora involved in anaerobic digestion of sewage sludge. Appl Environ Microbiol 66:2791–2796

Mitra SM, Guha Mazumder DN, Basu A, Block G, Haque R, Samanta S, Ghosh N, Smith MMH, von Ehrenstein OS, Smith AH (2004) Nutritional factors and susceptibility to arsenic-caused skin lesions in West Bengal, India. Environ Health Perspect 112:1104–1109

Mohapatra D, Mishra D, Rout M, Chaudhury GR (2007) Adsorption kinetics of natural dissolved organic matter and its impact on arsenic(V) leachability from arsenic loaded ferrihydrite and Al-ferrihydrite. J Environ Sci Health A Tox Hazard Subst Environ Eng 42:81–88

Mohapatra D, Mishra D, Chaudhury GR, Das RP (2008) Removal of arsenic from arsenic rich sludge by volatilization using anaerobic microorganisms treated with cow dung. Soil Sediment Contam 17:301–311

Mukhopadhyay R, Rosen BP (1998) *Saccharomyces cerevisiae ACR2* gene encodes an arsenate reductase. FEMS Microbiol Lett 168:127–136

Mukhopadhyay R, Rosen BP, Phung LT, Silver S (2002) Microbial arsenic: from geocycles to genes and enzymes. FEMS Microbiol Rev 26:311–325

Muller D, Médigue C, Koechler S et al (2007) A tale of two oxidation states: bacterial colonization of arsenic-rich environments. PLoS Genet 3:518–530

Murase J, Kimura M (1997) Anaerobic reoxidation of Mn^{2+}, Fe^{2+}, S^0 and S^{2-} in submerged paddy soils. Biol Fertil Soils 25:302–306

Nagvenkar GS, Ramaiah N (2010) Arsenite tolerance and biotransformation potential in estuarine bacteria. Ecotoxicology 19:604–613

Ng JC, Wang JP, Shraim A (2003) A global health problem caused by arsenic from natural sources. Chemosphere 52:1353–1359

Nies DH (1999) Microbial heavy-metal resistance. Appl Microbiol Biotechnol 51:730–750

Nies DH, Silver S (1995) Ion efflux systems involved in bacterial metal resistances. J Ind Microbiol 14:186–199

Ordonez E, Letek M, Valbuena N, Gil AJ, Mateos LM (2005) Analysis of genes involved in arsenic resistance in *Corynebacterium glutamicum* ATCC 13032. Appl Environ Microbiol 71:6206–6215

Oremland RS, Stolz JF (2003) The ecology of arsenic. Science 300:939–944

Oremland RS, Stolz JF (2005) Arsenic, microbes and contaminated aquifers. Trends Microbiol 13:45–48

Oremland RS, Dowdle PR, Hoeft S, Sharp JO, Schaefer JK, Miller LG, Switzer Blum J, Smith RL, Bloom NS, Wallschlaeger D (2000) Bacterial dissimilatory reduction of arsenate and sulfate in meromictic Mono Lake, California. Geochim Cosmochim Acta 64:3073–3084

Osman D, Cavet JS (2010) Bacterial metal-sensing proteins exemplified by ArsR-SmtB family repressors. Nat Prod Rep 27:668–680

Paez-Espino D, Tamales J, de Lorenzo V, Canovas D (2009) Microbial responses to environmental arsenic. Biometals 22:117–130

Pantsar-Kallio M, Korpela A (2000) Analysis of gaseous arsenic species and stability studies of arsine and trimethylarsine by gas-chromatography-mass spectrometry. Anal Chim Acta 410:65–70

Pennisi E (2010) What poison? Bacterium uses arsenic to build DNA and other molecules. Science 330:1302

Petänen T, Virta M, Karp M, Romantschuk M (2001) Construction and use broad host range mercury and arsenite sensor plasmids in the soil bacterium *Pseudomonas fluorescens OS8*. Microb Ecol 41:360–368

Plant JA, Kinniburgh DG, Smedley PL, Fordyce FM, Klinck BA (2004) Arsenic and selenium. In: Holland HD, Turekian KK (eds) Treatise on geochemistry, vol 9, Environmental geochemistry. Elsevier, Amsterdam, pp 17–66

Polemio M, Senesi N, Bufo SA (1982) Soil contamination by metals. A survey in industrial and rural areas of Southern Italy. Sci Total Environ 25:71–79

Pongratz W, Endler PC, Poitevin B, Kartnig T (1995) Effect of extremely diluted plant hormone on cell culture. Proc AAAS Ann Meeting, Atlanta

Porta D, Bullerjahn GS, Durham KA, Wilhelm SW, Twiss MR, McKay RML (2003) Physiological characterization of a *Synechococcus* sp. (cyanophyceae) strain PCC7942 iron-dependent bioreporter for freshwater environments. J Phycol 39:64–73

Purkayastha TJ (2011) Microbial remediation of arsenic contaminated soil. In: Sherrameti I, Varma A (eds) Detoxification of heavy metals, vol 30, Soil biology. Springer, Berlin, Heidelberg, pp 221–260

Qin J, Rosen BP, Zhang Y, Wang G, Franke S, Rensing C (2006) Arsenic detoxification and evolution of trimethylarsine gas by a microbial arsenite *S*-adenosylmethionine methyltransferase. Proc Natl Acad Sci U S A 103:2075–2080

Qin XS, GH Huang, Chakma A, Chen B, Zeng GM (2007) Simulation-based process optimization for surfactant - enhanced aquifer remediation at heterogeneous DNAPL-contaminated sites. Sci Total Environ 381:17–37

Rahmann MM, Chowdhury UK, Mukherjee SC, Mandal BK, Paul K, Lodh D, Biswas BK, Chanda CR, Basu GK, Saha KC, Roy S, Das R, Palit SK, Quamruzzaman Q, Chakraborti D (2001) Chronic arsenic toxicity in Bangladesh in West Bengal, India—a review commentary. J Toxicol Clin Toxicol 39:683–700

Rensing C, Rosen B (2009) Heavy metals cycle (arsenic, mercury, selenium, others). In: Schaeter M (ed) Encyclopedia of microbiology. Elsevier, Oxford, pp 205–219

Rensing C, Ghosh M, Rosen B (1999) Families of soft-metal-ion-transporting ATPases. J Bacteriol 181:5891–5897

Roberto F, Barnes J, Bruhn D (2002) Evaluation of a GFP reporter gene construct for environmental arsenic detection. Talanta 58:181–188

Robinson BH, Brooks RR, Howes AW, Kirman JH, Gregg PEH (1997) The potential of the high-biomass nickel hyperaccumulator *Berkheya coddii* for phytoremediation and phytomining. J Geochem Explor 60:115–126

Roos G, Loverix S, Brosens E, Van Belle K, Wyns L, Geerlings P, Messens J (2006) The activation of electrophile, nucleophile and leaving group during the reaction catalysed by pI258 arsenate reductase. Chembiochem 7:981–989

Rosen BP (1996) Bacterial resistance to heavy metals. J Biol Inorg Chem 1:273–277

Rosen BP (1999) Families of arsenic transporters. Trends Microbiol 7:207–212

Rosen BP (2002) Biochemistry of arsenic detoxification. FEBS Lett 529:86–92

Rosen BP, Weigel U, Karkaria C, Gangola P (1988) Molecular characterization of an anion pump. The *arsA* gene product is an arsenite (antimonate)-stimulated ATPase. J Biol Chem 263:3067–3070

Rosenberg HR, Gerdes RG, Chegwidden K (1977) Two systems for the uptake of phosphate in *E. coli*. J Bacteriol 131:505–511

Rosenzweig AC (2002) Metallochaperones: bind and deliver. Chem Biol 9:673–677

Ross S (1994) Toxic metals in soil-plant systems. Wiley, Chichester, UK

Rowland HAL, Pederick RL, Polya DA, Pancost RD, van Dongen BE, Gault AG, Vaughan DJ, Bryant C, Anderson B, Lloyd JR (2007) The control of organic matter on microbially mediated iron reduction and arsenic release in shallow alluvial aquifers, Cambodia. Geobiology 5:281–292

Ruan X, Bhattacharjee H, Rosen BP (2006) Cys-113 and Cys-422 form a high affinity metalloid binding site in the ArsA ATPase. J Biol Chem 281:9925–9934

Russeva E (1995) Speciation analysis—peculiarities and requirements. Anal Lab 4:143–148

Sauge-Merle S, Cuine S, Carrier P, Lecomte-Pradines C, Luu DT, Peltier G (2003) Enhanced toxic metal accumulation in engineered bacterial cells expressing Arabidopsis thaliana phytochelatin synthase. Appl Environ Microbiol 69:490–494

Saunders JA, Lee MK, Uddin A, Mohammed S, Wilkin RT, Fayek M, Korte NE (2005) Natural arsenic contamination of Holocene alluvial aquifers by linked tectonic, weathering, and microbial processes. Geochem Geophys Geosyst 6:1–7

Senn DB, Hemond HF (2004) Particulate arsenic and iron during anoxia in a eutrophic, urban lake. Environ Toxicol Chem 23:1610–1616

Sharples JM, Meharg AA, Chambers SM, Cairney JWG (2000) Mechanism of arsenate resistance in the ericoid mycorrhizal fungus *Hymenoscyphus ericae*. Plant Physiol 124:1327–1334

Shi W, Wu J, Rosen BP (1994) Identification of a putative metal binding site in a new family of metalloregulatory proteins. J Biol Chem 269:19826–19829

Shi W, Dong J, Scott RA, Ksenzenko MY, Rosen BP (1996) The role of arsenic-thiol interactions in metalloregulation of the *ars* operon. J Biol Chem 271:9291–9297

Silver S, Keach D (1982) Energy-dependent arsenate efflux: the mechanism of plasmid-mediated resistance. Proc Natl Acad Sci U S A 79:6114–6118

Silver S, Phung LT (1996) Bacterial heavy metal resistance: new surprises. Annu Rev Microbiol 50:753–789

Silver S, Phung LT (2005) Genes and enzymes involved in bacterial oxidation and reduction of inorganic arsenic. Appl Environ Microbiol 71:599–608

Silver S, Budd K, Leahy KM, Shaw WV, Hammond D, Novick RP, Willsky GR, Malamy MH, Rosenberg H (1981) Inducible plasmid-determined resistance to arsenate, arsenite, and antimony (III) in *Escherichia coli* and *Staphylococcus aureus*. J Bacteriol 146:983–996

Singh N, Ma LQ (2006) Arsenic speciation and arsenic phosphate distribution in arsenic hyperaccumulation *Pteris vittata* (L) and non-hyperaccumulator *Pteris ensiformis* (L). Environ Pollut 141:238–246

Singh S, Lee W, DaSilva NA, Mulchandani A, Chen W (2008a) Enhanced arsenic accumulation by engineered yeast cells expressing *Arabidopsis thaliana* phytochelatin synthase. Biotechnol Bioeng 99:333–340

Singh S, Mulchandani A, Chen W (2008b) Highly selective and rapid arsenic removal by metabolically engineered *Escherichia coli* cells expressing *Fucus vesiculosus* metallothionein. Appl Environ Microbiol 74:2924–2927

Smedley PL, Kinniburgh DG (2002) A review of the source, behavior and distribution of arsenic in natural waters. Appl Geochem 17:517–568

Sohrin Y, Matsui M, Kawashima M, Hojo M, Hasegawa H (1997) Arsenic biogeochemistry affected by eutrophication in lake Biwa, Japan. Environ Sci Technol 31:2712–2720

Song WY, Sohn EJ, Martinoia E, Lee YJ, Yang YY, Jasinski M, Forestier C, Hwang I, Lee Y (2003) Engineering tolerance and accumulation of lead and cadmium in transgenic plants. Nat Biotechnol 21:914–919

Sticher P, Jaspers MCM, Stemmler K, Harms H, Zehnder AJB, van der Meer JR (1997) Development and characterization of a whole-cell bioluminescent sensor for bioavailable middle-chain alkanes in contaminated groundwater samples. Appl Environ Microbiol 63:4053–4060

Stolz JF, Basu P, Santini JM, Oremland RS (2006) Arsenic and selenium in microbial metabolism. Annu Rev Microbiol 60:107–130

Strosnider H (2003) Whole-cell bacterial biosensors and the detection of bioavailable arsenic. Office of solid waste and emergency response technology innovation Environmental Protection Agency, Washington DC

Sun Y, Wong MD, Rosen BP (2001) Role of cysteinyl residues in sensing Pb(II), Cd(II), and Zn(II) by the plasmid pI258 CadC repressor. J Biol Chem 276:14955–14960

Takeuchi M, Kawahata H, Gupta LP, Kita N, Morishita Y, Onoc Y, Komai T (2007) Arsenic resistance and removal by marine and non-marine bacteria. J Biotechnol 127:434–442

Tauriainen S, Karp M, Virta M (1997) Recombinant luminescent bacteria for measuring bioavailable arsenite and antimonite. Appl Environ Microbiol 63:4456–4461

Taurianen S, Virta M, Chang W, Karp M (1999) Measurement of firefly luciferase reporter gene activity from cells and lysates using *Escherichia coli* arsenate and mercury sensors. Anal Biochem 272:191–198

Tisa LS, Rosen BP (1990) Molecular characterization of an anion pump. The ArsB protein is the membrane anchor for the ArsA protein. J Biol Chem 265:190–194

Trotta A, Falaschi P, Cornara L, Minganti V, Fusconi A, Drava G, Berta G (2006) Arbuscular mycorrhizae increase the arsenic translocation factor in the As hyperaccumulating fern *Pteris vittata* L. Chemosphere 65:74–81

Tsai SL, Singh S, Chen W (2009) Arsenic metabolism by microbes in nature and the impact on arsenic remediation. Curr Opin Biotechnol 20:1–9

Tu C, Ma LQ, Bondada B (2002) Arsenic accumulation in the hyperaccumulator chinese brake fern (*Pteris vittata* L.) and its utilization potential for phytoremediation. J Environ Qual 31:1671–1675

Tu S, Ma LQ, MacDonald GE, Bondada B (2003) Effects of arsenic species and phosphorus on arsenic absorption, arsenate reduction and thiol formation in excised parts of *Pteris vittata* L. Environ Exp Bot 51:121–131

Tu S, Ma LQ, Fayiga AO, Zillioux EJ (2004) Phytoremediation of arsenic-contaminated groundwater by the arsenic hyperaccumulating fern *Pteris vittata* L. Int J Phytoremediation 6:35–47

Tuffin IM, Hector SB, Deane SM, Rawling DE (2006) Resistance determinants of highly arsenic-resistant strain of *Leptospirillum ferriphilum* isolated from a commercial biooxidation tank. Appl Environ Microbiol 72:2247–2253

Turpeinen R (2002) Interactions between metals, microbes and plants: bioremediation of arsenic and lead contaminated soils. A dissertation in environmental ecology, Faculty of Science, University of Helsinki, Neopoli, Lahti

Turpeinen R, Pantsar-Kallio M, Haggblom M, Kairesalo T (1999) Influence of microbes on the mobilization, toxicity and biomethylation of arsenic in soil. Sci Total Environ 236:173–180

Turpeinen R, Virta M, Haggblom M (2003) Analysis of arsenic bioavailability in contaminated soils. Environ Toxicol Chem 22:1–6

Waldron KJ, Robinson NJ (2009) How do bacterial cells ensure that metalloproteins get the correct metal. Nat Rev Microbiol 7:25–35

Wang S, Mulligan CN (2006) Natural attenuation processes for remediation of arsenic contaminated soils and groundwater. J Hazard Mater B138:459–470

Wang L, Chen S, Xiao X, Huang X, You D, Zhou X, Deng Z (2006) *arsRBOCT* arsenic resistance system encoded by linear plasmid pHZ227 in *Streptomyces* sp. strain FR-008. Appl Environ Microbiol 72:3738–3742

Welch AH, Lico MS (1998) Factors controlling As and U in shallow ground water, southern Carson Dessert, Nevada. Appl Geochem 13:521–539

WHO (2001) Arsenic in drinking water, Fact sheet No. 210, Revised Edition (http://www.who.int/ mediacentre/factsheet/fs21)

Wolfe-Simon F, Blum JS, Kulp TR, Gordon GW, Hoeft SE, Pett-Ridge J, Stolz JF, Webb SM, Weber PK, Davis PCW, Anbar AD, Oremland RS (2011) A bacterium that can grow by using arsenic instead of phosphorus. Science 332:1163–1166

Woolson EA (1977) Generation of alkylarsines from soil. Weed Sci 25:412–416

Wu J, Tisa LS, Rosen BP (1992) Membrane topology of the ArsB protein, the membrane subunit of an anion-translocating ATPase. J Biol Chem 267:12570–12576

Wu B, Song J, Beitz E (2010) Novel channel enzyme fusion proteins confer arsenate resistance. J Biol Chem 285:40081–40087

Wuilloud RG, Altamirano JC, Smichowski PC, Heitkemper DT (2006) Investigation of arsenic speciation in algae of the Antarctic region by HPLC-ICP-MS and HPLC-ESI-Ion Trap MS. J Anal At Spectrom 21:1214–1223

Xu C, Shi W, Rosen BP (1996) The chromosomal *arsR* gene of *E. coli* encodes a trans-acting metalloregulatory protein. J Biol Chem 271:2427–2432

Yang HC, Cheng J, Finan TM, Rosen BP, Bhattacharjee H (2005) Novel pathway for arsenic detoxification in the legume symbiont *Sinorhizobium meliloti*. J Bacteriol 187:6991–6997

Yang J, Rawat S, Stemmler TL, Rosen BP (2010) Arsenic binding and transfer by the ArsD As(III) metallochaperone. Biochemistry 49:3658–3666

Zawadzka AM, Crawford RL, Paszczynski AJ (2007) Pyridine-2,6-bis(thiocarboxylic acid) produced by *Pseudomonas stutzeri* KC reduces chromium(VI) and precipitates mercury, cadmium, lead and arsenic. Biometals 20:145–158

Zegers I, Martins JC, Willem R, Wyns L, Messens J (2001) Arsenate reductase from *S. Aureus* plasmid pI258 is a phosphatase drafted for redox duty. Nat Struct Biol 8:843–847

Zhou T, Rosen BP (1997) Tryptophan fluorescence reports nucleotide-induced conformational changes in a domain of the ArsA ATPase. J Biol Chem 272:19731–19737

Zhou T, Radaev S, Rosen BP, Gatti DL (2000) Structure of the ArsA ATPase: the catalytic subunit of a heavy metal resistance pump. EMBO J 19:1–8

Zhou Y, Messier N, Ouellette M, Rosen BP, Mukhopadhyay R (2004) *Leishmania major* LmACR2 is a pentavalent antimony reductase that confers sensitivity to the drug pentostam. J Biol Chem 279:37445–37451

Zobrist J, Dowdle PR, Davis JA, Oremland RS (2000) Mobilization of arsenite by dissimilatory reduction of arsenate. Environ Sci Technol 34:4747–4753

Zuber P (2004) Spx-RNA polymerase interaction and global transcriptional control during oxidative stress. J Bacteriol 186:1911–1918

Chemical Behavior of Phthalates Under Abiotic Conditions in Landfills

Jingyu Huang, Philip N. Nkrumah, Yi Li, and Gloria Appiah-Sefah

Contents

1 Introduction.. 39
2 General Description of the Phthalates.. 40
   2.1 Synthesis (Esterification Reaction).. 40
   2.2 Physical/Chemical Properties 41
   2.3 Adverse Effects 42
3 Occurrence of Phthalate Esters in Landfills.. 42
4 Decomposition of the Phthalates.. 43
   4.1 Abiotic Degradation 43
   4.2 Biodegradation 47
5 Summary 48
References.. 49

1 Introduction

Phthalates or phthalic acid esters (PAEs) are diesters of phthalic anhydride. They are synthesized from an esterification reaction between phthalic anhydride and oxo alcohols (ECOBILAN 2001). Phthalates are usually used as plasticizers to enhance the flexibility of materials and their technical properties. Mersiowsky et al. (2001) reported that the phthalates serve as plasticizers for approximately 93% of the polyvinyl chloride (PVC) polymer that is produced. In addition, they are also used in cosmetics, in fragrances, as pesticide carriers, in insect repellants, and are found in vinyl floorings, wall coverings, cables, tubing, hoses, upholstery, films, paints, adhesives, and inks, among other products (ECPI 1994; Schierow and Lee 2008).

J. Huang • P.N. Nkrumah (✉) • Y. Li • G. Appiah-Sefah
Ministry of Education, Key Laboratory of Integrated Regulation and Resource
Development on Shallow Lakes, College of Environment, Hohai University,
No. 1 Xikang Road, Nanjing 210098, China
e-mail: philiponti1209@yahoo.com

D.M. Whitacre (ed.), *Reviews of Environmental Contamination and Toxicology*,   39
Reviews of Environmental Contamination and Toxicology 224,
DOI 10.1007/978-1-4614-5882-1_2, © Springer Science+Business Media New York 2013

The annual worldwide production of PAEs exceeds five million tons (Mackintosh et al. 2006).

Because the phthalate esters are not covalently bound to the polymer of which they are a component, they are able to migrate to the surface of the polymer matrix where they may be lost by a variety of physical processes (Stanley et al. 2003). Fromme et al. (2002), Fernandez et al. (2007), and Zeng et al. (2009) have noted that phthalates, because of their fugitive nature and widespread use, are commonly detected in air, water, sediment/soil, and biota including human tissue. Bauer et al. (1998) observed that large amounts of phthalic acid esters are often leached from the plastics that are dumped at municipal landfills and enumerated that these pollutants are harmful to microorganisms and accumulate in natural bodies of water. Residues of these contaminants ultimately become widely distributed within aqueous systems such as rivers, lakes, and groundwater and thereby exert a noticeable influence on the environment.

Evidence procured from epidemiological studies with humans shows that phthalates induce adverse health effects that include disorders of the male reproductive tract, breast and testicular cancers, and neuroendocrine system disruption (Sharpe and Shakkebaek 1993; Matsumoto et al. 2008; IHCP 2008; Huang et al. 2009).

When the phthalates are released into the environment, they interact with environmental media in ways that change their chemical behavior. Staples et al. (2000) outlined the various processes to which the phthalates are subjected when released into the environment. These processes include hydrolysis, photolysis, and biodegradation. The relevance of these processes is observed to be dependent upon the properties of the phthalate ester and the ambient physical and chemical conditions that exist in any particular landfill zone (Gächter and Müller 1990; Domininghaus 1998).

In this review, we focus on the chemical transformation that the phthalate esters undergo when they are released into the lower landfill layers. Specifically, we aim to present the possible ways in which this chemical is transformed under the prevailing conditions that exist in lower landfill layers, which include high temperature, presence of chemical catalysts, and fluctuating pH levels.

2   General Description of the Phthalates

2.1   Synthesis (Esterification Reaction)

PAEs are diesters of phthalic anhydride and have the common chemical structure shown in Fig. 1. Their synthesis involves reacting phthalic anhydride with oxo alcohols to form esters. According to ECOBILAN (2001), PAEs differ from the nature and length of the oxo alcohols (C_1 to C_{13}) from which they are made.

The PAEs are produced by a final esterification reaction consisting of two steps (ECOBILAN 2001). The first step involves the alcoholysis of phthalic anhydride to form the monoester and secondly, the conversion of the monoester to a diester which

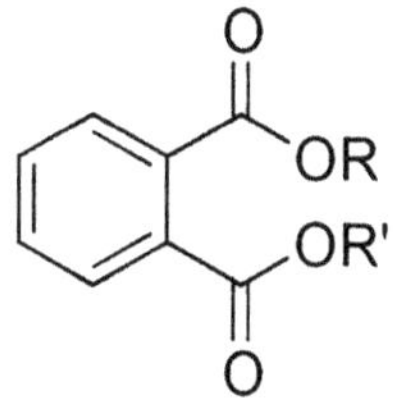

Fig. 1 Phthalate structure
(*R* and *R'* represent alkyl side
chains which may be branched
and contain oxygen)

is a reversible reaction. According to Skrzypek et al. (2008), this first step is fast and irreversible, whilst the second step is slow and usually requires a catalyst.

For example, the synthesis of di(2-ethylhexyl) phthalate (DEHP), as performed by Skrzypek et al. (2008), results in the monoester being formed on the dissolution of phthalic anhydride in 2-ethylhexyl alcohol at temperatures of 320 K–360 K, and this occurs without a catalyst. These authors proposed that less aggressive catalysts, such as *p*-toluenesulfonic acid or methane sulfonic acid, could be employed to speed the reaction rate occurring in the second step, rather than by using sulfuric acid. This strong acid has been employed, both in laboratory and industrial practice, despite its having the disadvantage of generating undesirable amounts of by-products that result in color change or blushing of the product.

The rate of the above reaction is dependent on the catalyst chosen and the reaction temperature (Skrzypek et al. 2008).

2.2 Physical/Chemical Properties

Liang et al. (2008) disclosed the basic chemical structure of phthalates to be benzene dicarboxylic acid, with two side chains that may be comprised of chemical groups that include alkyl, benzyl, phenyl, cycloalkyl, or alkoxy.

Phthalate esters are liquids at typical environmental temperatures. Melting points for these esters are between 5.5°C and −58°C, and boiling points are between 230 and 486°C (Staples et al. 1997; Cousins et al. 2003). Stanley et al. (2003) established that the water solubility of the alkyl phthalate ester generally varies inversely with the length of the alkyl side chain. Dimethyl phthalate (DMP) is the most hydrophilic and water soluble of the esters. The C_{10}, C_{11}, and C_{13} esters are the most hydrophobic and least water soluble (<0.001 mg/L).

The sorption of phthalate esters to soil, sediment, or suspended solids is partially governed by their relative hydrophobicity (Staples et al. 1997). Cousins et al. (2003) asserted that phthalates have high octanol–air partition coefficient (K_{OA}) values, suggesting that they will be appreciably sorbed to aerosols, soils, and vegetation.

According to Woodward (1988), most dialkyl phthalates are soluble in common organic solvents such as benzene, toluene, xylene, diethyl ether, chloroform, and petroleum ether.

It is noteworthy that the degradation rate of the phthalates is dependent on their molecular weight; those with longer alkyl side chains tend to have longer half-lives in a given environmental media (Harris and Sumpter 2001).

2.3   Adverse Effects

Their extensive use has made the PAEs ubiquitous environmental pollutants, and they are released to the environment from both diffuse and point sources (Furtmann 1996; Chatterjee and Karlovsky 2010). These pollutants may harm microorganisms and may reach and accumulate in natural bodies of water to ultimately become widely distributed in rivers, lakes, and groundwater, where they may exert adverse environmental effects (Bauer et al. 1998).

Moreover, the phthalates and their metabolites may pose potentially harmful effects on humans and the environment because of their hepatotoxic, teratogenic, and carcinogenic characteristics (Matsumoto et al. 2008).

The US EPA (2001) reported that plasticizers are one of the most prominent classes of endocrine-disrupting chemicals (EDCs). As EDCs, their adverse effects on reproduction and development have been demonstrated in animal models. Foster (2006), Ghisari and Bonefeld-Jorgensen (2009), Gray et al. (2009), and Strac (2009) have indicated that these compounds decrease birth weight and survival of offspring, induce testicular atrophy and malformation, impair spermatogenesis and reduce sperm counts, and shorten the anogenital distance in rodents and other animals. Most of these effects are thought to be caused by mimicking or antagonizing the actions of endogenous steroid hormones (Foster 2006; Ghisari and Bonefeld-Jorgensen 2009; Gray et al. 2009; Nagao et al. 2000; Shen et al. 2009). Zhang et al. (2011b) evaluated the estrogen agonist/antagonist properties of phthalates such as of dibenzyl phthalate (DBzP). In their study, these properties were predicted by molecular docking and confirmed by yeast estrogen screen (YES) and immature mouse uterotrophic assays. These authors confirmed the effects of phthalates on estrogens. According to Gültekin and Ince (2007), despite their low concentration in the aquatic environment, such labeled hormone-like chemicals may be "hazardous," because even trace amounts may produce estrogenic-like activity.

3   Occurrence of Phthalate Esters in Landfills

Phthalates have been detected in various environmental media including air (Wensing et al. 2005), soils, sediments, landfill leachates (Schwarzbauer et al. 2002; Zheng et al. 2007), and in natural waters as a result of their production, usage, and disposal in plastics (Chao and Cheng 2007). PAEs are produced in large quantities and are used in a wide variety of products that are ultimately disposed of in landfills. The most common use of phthalic acid diesters is as plasticizers in PVC plastic (Gomez-Hens and Aguilar-Caballos 2003).

Stanley et al. (2003) and Liang et al. (2008) have both indicated that phthalate esters are not covalently bound to the polymer of which they are a component and therefore migrate to the surface of the polymer matrix where they may be lost to the environment. Large amounts of phthalic acid esters are often leached from the plastics that are dumped at municipal landfills. Bauer et al. (1998) and Jonsson et al. (2003) specifically observed that PAEs, not being chemically bound to the plastic structure, may be leached into soils by water percolating through landfills.

Christensen et al. (2001) described the composition of landfill leachate as including dissolved organic matter, inorganic macro components, heavy metals, and Xenobiotic Organic Compounds (XOCs). The presence of water, acid, base, and heavy metals, among other substances in the landfill, facilitates the abiotic transformation of phthalates at lower layers of a municipal landfill. For example, heavy metals such as nickel serve as catalysts in the hydrogenation of dialkyl ester of phthalic acid to the corresponding dialkyl hexahydrophthalates (Amend 1935).

The temperatures usually encountered in a typical landfill body are in the range of 18–55°C with the average at 35°C (Dohmann 1997). However, Schwarzbauer et al. (2006) observed temperature values as high as 70°C within the lower layers of a landfill.

4 Decomposition of the Phthalates

Generally, the major processes that aid in the environmental transformation of phthalates are biodegradation, photolysis, and hydrolysis. Cousins et al. (2003) reported that biodegradation is the dominant loss process for the phthalate esters in all media, except the atmosphere, wherein they are likely to be susceptible to rapid photooxidation by hydroxyl radicals. Chung and Chen (2009) suggested that microbial degradation and abiotic degradation occur by different mechanisms.

Below, we describe the chemical behavior that phthalate esters generally display when they are released into landfill compartments. We assert that hydrolysis is the dominant abiotic process that affects the transformation of the phthalates within the lower landfill layers. Schwarzbauer et al. (2006) observed that as landfill depth increases the propensity for ongoing hydrolysis increases.

4.1 Abiotic Degradation

Few studies have been performed to ascertain the chemical transformation that occurs for phthalates under the prevailing conditions present in lower landfill layers; studies that have been performed show that abiotic processes, such as hydrolysis, generally proceed at a slow rate (Schwartzenbach et al. 1992; Yan et al. 1995; Staples et al. 1997; Schwarzbauer et al. 2006). Lertsirisopon et al. (2009) also bemoaned that limited information was available on abiotic processes in landfills.

However, although some microbial action may occur, it may be negligible in lower layers of a landfill because of the high temperatures and pressures within this zone. According to Miller (1992), temperatures in excess of 60°C within a medium reduce microbial activity in that medium, and above this temperature, microbial activity declines. The temperatures that exist at lower waste layers of a landfill can be as high as 70°C (Schwarzbauer et al. 2006).

Local physical and chemical conditions greatly affect the rate of abiotic transformation of the phthalates (Gächter and Müller 1990; Dominghaus 1998). Sayyed et al. (2010) reported that the rate of degradation was influenced by the following factors: temperature, pressure, substrate concentration, oxidant concentration, composition of reaction mixture, and presence of chemical catalysts.

Hydrolysis and photolysis are the major abiotic transformation processes to which the phthalate esters are subjected when they are released into the environment (Staples et al. 1997). Hydrogenation of the phthalates is another transformation process that also occurs under certain environmental conditions (Schwarzbauer et al. 2006).

Below, we address, in greater detail, the key abiotic processes that are responsible for transforming phthalates in landfills.

4.1.1 Hydrolysis

Hydrolysis is a chemical transformation process in which an organic compound, RX, reacts with water, forming a new carbon–oxygen bond, and the cleaving of the carbon-X bond in the original molecule (Hilal 2006). Hydrolysis is one of the most common reactions controlling abiotic degradation and is therefore one of the main degradation paths for xenobiotic substances (OECD 1981). Generally, the hydrolysis of an ester forms a carboxylic acid and an alcohol.

Staples et al. (1997) reported that the phthalate esters are susceptible to hydrolysis, but at rather slow rates. These authors noted that phthalate esters can undergo two hydrolytic steps, producing first the monoester and one free alcohol moiety and a second hydrolytic step creating phthalic acid and a second alcohol. Schwarzbauer et al. (2006) described the typical abiotic transformation process that occurs for the phthalate esters at lower layers of a municipal landfill as follows:

- Initially, the phthalate diesters were converted to the corresponding monoesters via hydrolysis. These monoesters were then converted to phthalic acid (Fig. 2).
- The generated phthalic acid was subjected to extended hydrogenation, which led to the formation of hydrogenation product (1,2-cyclohexane dicarboxylic acid). These authors observed that the concentration of the hydrogenation product in the layers below 50 m remained at a high level; as a result, ongoing generation of phthalic acid was no longer interfered with by consecutive hydrogenation processes.
- Finally, phthalic acid was observed to accumulate in the lower waste layers.

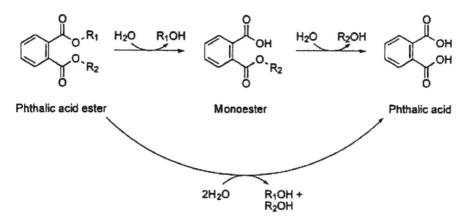

Fig. 2 Transformation of phthalates via hydrolysis (Adapted from Staples et al. 1997; Mersiowsky et al. 1999; Schwarzbauer et al. 2006; Shibata et al. 2007)

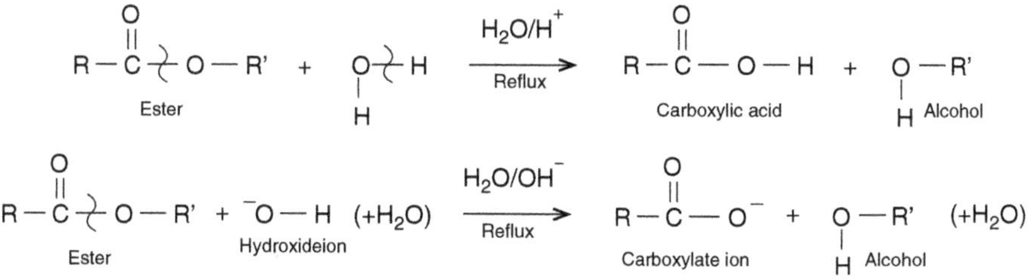

Fig. 3 Acid and base hydrolysis (Adapted from Sykes 1975; Wolfe et al. 1980; Harris 1982; Hilal 2006)

Schwarzbauer et al. (2006) further pointed out that the above-described processes were affected by the temperature gradient. It is also noteworthy that the hydrogenation process that occurs in deep waste layers may result from either higher temperature or the restriction of reactions involving the free aromatic acid.

Hence, the prevailing conditions existing at lower landfill layers are favorable to hydrolysis and represent the main route by which the phthalates are transformed within this landfill zone.

The effect of pH on the hydrolysis of the phthalates: Mabey and Mill (1978) indicated that hydrolysis is basically a function of pH and temperature. Notably, pH affects reaction rates through catalysis. After shifts in pH, reaction rates of esters may change some orders of magnitude from catalysis (Kirby 1972). Esters are much less electrophilic and do not react with water at pH 7 to any appreciable extent. The author noted, however, that the rate of ester hydrolysis is substantially increased under acidic or basic conditions. Ester hydrolysis may be either acid- or base-catalyzed, with metal ions, anions, or organic materials serving as catalysts (Sykes 1975; Harris 1982; US EPA 1996). Figure 3 illustrates both acid- and base-catalyzed hydrolysis reactions.

Hilal (2006) asserted that carboxylic acid esters undergo hydrolysis through three different mechanisms: base-, acid-, and general base-catalyzed (neutral) hydrolysis. For instance, Patnaik et al. (2001) conducted a study to investigate base-catalyzed hydrolysis of the phthalates and ammonium hydroxide present in landfill. In this study, DMP and diethyl phthalate (DEP) reactions with ammonium hydroxide were carried out under alkaline conditions in the pH range 8.5–9.5 at ambient temperature and at a concentration range of 5–20 mg/L. Wolfe et al. (1980) also observed that PAEs were hydrolyzed to the corresponding monoesters under acidic conditions.

It is noteworthy that the phthalate esters are hydrolyzed at negligible rates at a neutral pH (Staples et al. 1997). Xu et al. (2008) also confirmed this assertion. It is important to note that acid hydrolysis of phthalate esters occurs, but is estimated to be four orders of magnitude slower than alkaline hydrolysis rate constants (Mabey et al. 1982).

Temperature effects on the hydrolysis of the phthalates: As already mentioned, hydrolysis is a function of temperature (Mabey and Mill 1978; Hilal et al. 2003; Hilal 2006). Moreover, it is a known fact that landfill temperature increases with depth. Therefore, increasing temperature with depth in a landfill may well enhance the rate of processes responsible for transformation. Schwarzbauer et al. (2006) observed that the transformation of phthalate-based plasticizers involves a consecutive order of hydrolysis and hydrogenation steps that is strongly influenced by transfer processes along the depth profile of a given landfill.

As the temperature increases, the steric effect decreases. The reason for this is attributed to the fact that as the temperature increases, the reactants tend to mimic gas-phase structure, producing minimal or null steric effect for ester hydrolysis.

4.1.2 Photocatalytic Oxidation

Phthalates are photocatalytically transformed through two different reaction mechanisms. The first one is via direct ozonation by ozone molecules, primarily of specific functional groups (double bonds, nucleophilic positions). The second is a radical oxidation by highly oxidative free radicals such as hydroxyl free radicals (OH·), which are generated from the decomposition of ozone in an aqueous solution (Gurol and Singer 1982; Hoffmann et al. 1995). This radical oxidation is nonselective and vigorous. According to Oh et al. (2006) and Zhang et al. (2011a), the reason ultraviolet (UV) radiation is introduced into the ozonation process is to enhance ozone decomposition, yielding more free radicals to achieve greater oxidation.

Chen et al. (2011) proposed a mechanism for the degradation of DMP during catalytic ozonation with a TiO_2/Al_2O_3 catalyst. These authors observed that the hydroxylation of DMP to 3-hydroxy dimethyl phthalate was the first ozonated step. The hydroxylation of DMP was followed by oxidative cleavage of the benzene ring; this carboxylation forms 3,4-dimethyloxycarbonyl 2,4-hexadienedioic acid. Further oxidation resulted in the intermediates 3-hydroxy3-formyl propanoic acid, 2-hydroxypentanedioic acid, and 2,3-dihydroxypentanedioic acid. Catalytic ozonation then preferably cleaves the benzene ring at the ortho position of the methyloxycarbonyl group to generate 5,6-dimethyloxycarbonyl 2,4-hexadieneoic acid on the surface of the TiO_2/Al_2O_3 catalyst. The intermediates that resulted included 2-hydroxy dimethyl succinate, 2,4-hexadienedioic acid, and 4-hydroxy 2-pentenoic acid.

4.1.3 Photolysis

The most important photodegradation reactions that occur in the atmosphere for organic chemical pollutants generally involve hydroxyl radicals (Atkinson 1988; Kelly et al. 1994). Staples et al. (1997) asserted that aqueous photolysis results from absorption of the UV spectrum (viz., in the region of 290–400 nm) of sunlight. According to these authors, photolysis occurs through mechanisms that either involve direct absorption of UV radiation by the chemical, or indirectly, when UV radiation is absorbed by natural substances (e.g., water), after which activated species are formed (e.g., singlet oxygen or hydroxyl radicals) that then react with phthalate esters.

4.2 Biodegradation

Within the upper landfill layers, the presence of microorganisms greatly advances the degradation of the phthalates, which prevails as the dominant transformation route within this zone (Ejlertsson et al. 1996; Staples et al. 1997; Peterson and Staples 2003; Di Gennaro et al. 2005). The biodegradation of the phthalates has been extensively studied and well documented (Elder and Kelly 1994; Ejlertsson et al. 1996; Staples et al. 1997; Kleerebezem et al. 1999; Jianlong et al. 2000; Juneson et al. 2001; Cousins et al. 2003; Liu and Chi 2003; Peterson and Staples 2003; Chang et al. 2004; Di Gennaro et al. 2005).

Generally, the biodegradation pathway for the phthalates consists of primary biodegradation from phthalate diesters (PDEs) to phthalate monoesters (PMEs) and then to phthalic acid (PA) and ultimate biodegradation of PA to CO_2 and/or CH_4 (Staples et al. 1997; Jianlong et al. 2000). This degradation process occurs under both aerobic and anaerobic conditions (Cousins et al. 2003).

Peterson and Staples (2003) reported that PA is degraded aerobically via hydroxylation and decarboxylation to give protocatechuic acid, which is further degraded to carbon dioxide via either *ortho* or *meta* cleavage of the aromatic ring. In contrast, PA is degraded anaerobically via decarboxylation to benzoic acid, which is further degraded through ring saturation to hydrogen, carbon dioxide, and acetate (Elder and Kelly 1994; Kleerebezem et al. 1999; Liu and Chi 2003).

Jianlong et al. (2000) and Chang et al. (2004) pointed out that phthalates with shorter ester chains, like DMP, DEP, dibutyl phthalate (DBP), diphenyl phthalate (DPP), and butyl benzyl phthalate (BBP), can be readily biodegraded and mineralized, whereas phthalates with longer ester chains, such as dihexyl phthalate (DHP) and DEHP, are less susceptible to biodegradation.

5   Summary

The phthalates comprise a family of phthalic acid esters that are used primarily as plasticizers in polymeric materials to impart flexibility during the manufacturing process and to the end product. It is estimated that the annual worldwide production of phthalate esters exceeds five million tons. Plasticizers are one of the most prominent classes of chemicals, but unfortunately, they possess endocrine-disrupting chemical properties. As endocrine-disrupting chemicals, plasticizers have produced adverse developmental and reproductive effects in mammalian animal models. Phthalates are easily transported into the environment during manufacture, disposal, and leaching from plastic materials, because they are not covalently bound to the plastics of which they are a component. Because of their fugitive nature and widespread use, the phthalates are commonly detected in air, water, sediment/soil, and biota, including human tissue. Large amounts of phthalic acid esters are often leached from the plastics that are dumped at municipal landfills.

Phthalate esters undergo chemical changes when released into the environment. The primary processes by which they are transformed include hydrolysis, photolysis, and biodegradation. It is noteworthy that all of these degradation processes are greatly influenced by the local physical and chemical conditions. Hence, in the present review, we have sought to ascertain from the literature how the phthalate esters undergo transformation when they are released into lower landfill layers.

Within the upper landfill layers, biodegradation prevails as the major degradation mechanism by which the phthalates are dissipated. Generally, biodegradation pathways for the phthalates consist of primary biodegradation from phthalate diesters to phthalate monoesters, then to phthalic acid, and ultimately biodegradation of phthalic acid to form CO_2 and/or CH_4.

We have noted that the phthalate esters are also degraded through abiotic means, which proceeds via both hydrolysis and photolysis. Photodegradation generally involves reactions of the phthalates in the atmosphere with hydroxyl radicals. The hydrolysis of phthalate diesters produces the corresponding monoesters, which are subsequently converted to phthalic acid. Phthalic acid has been observed to accumulate within landfill zones where phthalate contamination exists.

Hydrolysis is usually not an important fate process for phthalate esters in the environment, including in upper landfill layers. However, the conditions prevalent at lower landfill layers are generally suitable for phthalate transformation via hydrolysis. The conditions in this zone include high temperatures and pressures, presence of chemical catalysts, as well as wide pH fluctuations. Such conditions foster hydrolysis that may be either acid- or base-catalyzed by metal ions, anions, or organic materials catalysts. In addition, research indicates that the propensity for ongoing hydrolysis increases as landfill depth increases.

We can be emphatic in asserting that hydrolysis of phthalate esters in lower landfill layers is the dominant process for transforming these esters; in contrast, biodegradation is the predominant process in the upper landfill layers.

We recommend that future research be performed to expand the understanding of what influence each reaction condition (high temperature, presence of chemical catalysts, etc.) has on the rate of chemical transformation of the phthalates in lower landfill zones. We also recommend that the combined effects of all conditions on the rate of chemical transformation at lower landfill layers be assessed for the phthalates. Such research could be achieved under simulated conditions.

Acknowledgment We are extremely grateful to Dr. Dave Whitacre, RECT Editor, for his excellent comments and editing of the manuscript. The authors appreciate the input made by Subhankar Chatterjee and other anonymous reviewers. We would also like to thank Jennifer Abena Kwofie for her contribution.

References

Amend WJ (1935) Hydrogenation of alkyl phthalates. United States Patent Office. http://ip.com/patent/US2070770. Accessed 21 May 2012

Atkinson R (1988) Estimation of Gas-Phase hydroxyl radical rate constants for organic chemicals. Environ Toxicol Chem 7(6):435–442

Bauer MJ, Herrmann R, Martin A, Zellmann H (1998) Chemodynamics, transport behavior and treatment of phthalic acid esters in municipal landfill leachates. Water Sci Technol 38:185–192

Chang BV, Yang CM, Cheng CH, Yuan SY (2004) Biodegradation of phthalate esters by two bacteria strains. Chemosphere 55:533–538

Chao WL, Cheng CY (2007) Effect of introduced phthalate-degrading bacteria on the diversity of indigenous bacterial communities during di(2-ethylhexyl)phthalate (DEHP) degradation in a soil microcosm. Chemosphere 67:482–488

Chatterjee S, Karlovsky P (2010) Removal of the endocrine disrupter butyl benzyl phthalate from the environment. Appl Microbiol Biotechnol 87(1):61–73

Chen Y, Hsieh D, Shang N (2011) Efficient mineralization of dimethyl phthalate by catalytic ozonation using TiO_2/Al_2O_3 catalyst. J Hazard Mater 192:1017–1025

Christensen TH, Kjeldsen P, Bjerg PL, Jensen DL, Christensen JB, Baun A, Albrechtsen H, Heron G (2001) Biogeochemistry of landfill leachate plumes. Appl Geochem 16:659–718

Chung Y, Chen C (2009) Degradation of Di-(2-ethylhexyl)phthalate (DEHP) by TiO_2 photocatalysis. Water Air Soil Pollut 200:191–198

Cousins IT, Mackay D, Parkerton TF (2003) Physical-chemical properties and evaluative fate modelling of phthalate esters. In: Charles AS (ed) The handbook of environmental chemistry, Vol. 3, Part Q. Springer, New York, pp 57–84

Di Gennaro P, Collina E, Franzetti A, Lasagni M, Luridiana A, Pitea D (2005) Bioremediation of diethylhexyl phthalate contaminated soil: a feasibility study in slurry-and solid-phase reactors. Environ Sci Technol 39:325–330

Dohmann T (1997) Emission behaviour of pollutants. Report to the German Ministry of Research (BMBF) by Institut für Siedlungswasserwirtschaft, Rheinisch-Westfä -lische Technische Hochschule Aachen

Domininghaus H (1998) Plastics and their properties. Int J ChemTech Res 2:1

ECOBILAN (2001) Eco-profile of high volume commodity phthalate esters (DEHP/DINP/DIDP). European Council for Plasticisers & Intermediates (ECPI), Brussels

ECPI (European Council for Plasticisers and Intermediates) (1994) Phthalate esters used in PVC—assessment of the release, occurrence and possible effects of plasticizers in the environment. ECPI, Brussels

Ejlertsson J, Meyerson U, Svensson BH (1996) Anaerobic degradation of phthalic acid esters during digestion of municipal solid waste under landfilling conditions. Biodegradation 7:345–352

Elder DJE, Kelly DJ (1994) The bacterial degradation of benzoic-acid and benzenoid compounds under anaerobic conditions—unifying trends and new perspectives. FEMS Microbiol Rev 13:441–468

Fernandez MP, Ikonomou MG, Buchanan I (2007) An assessment of estrogenic organic contaminants in Canadian wastewaters. Sci Total Environ 373:250–269

Foster PM (2006) Disruption of reproductive development in male rat offspring following in utero exposure to phthalate esters. Int J Androl 29:140–147, discussion 181–185

Fromme H, Kücher T, Otto T, Pilz K, Müller J, Wenzel A (2002) Occurrence of phthalates and bisphenol A and F in the environment. Water Res 36:1429–1438

Furtmann K (1996) Phthalates in the aquatic environment. European Council for Plasticisers & Intermediates (ECPI), Brussels

Gächter R, Müller H (1990) Handbook of plastics additives, 3rd edn. Carl Hanser, Munich. ISBN 3-446-15627-5

Ghisari M, Bonefeld-Jorgensen EC (2009) Effects of plasticizers and their mixtures on estrogen receptor and thyroid hormone functions. Toxicol Lett 189:67–77

Gomez-Hens A, Aguilar-Caballos MP (2003) Social and economic interest in the control of phthalic acid esters. Trends Anal Chem 22:847–857

Gray LE, Barlow NJ, Howdeshell KL, Ostby JS, Furr JR, Gray CL (2009) Transgenerational effects of di (2-ethylhexyl) phthalate in the male CRL:CD(SD) rat: added value of assessing multiple offspring per litter. Toxicol Sci 110:411–425

Gültekin I, Ince NH (2007) Synthetic endocrine disruptors in the environment and water remediation by advanced oxidation processes. J Environ Manage 85(4):816–832

Gurol MD, Singer PC (1982) Kinetic of ozone decomposition: a dynamic approach. Environ Sci Technol 16:377–383

Harris JC (1982) Rate of hydrolysis. In: Lyman WJ, Reehl WF, Rosenblart DH (eds) Handbook of chemical property estimation methods, chapter 7. McGraw-Hill, New York

Harris CA, Sumpter JP (2001) The endocrine disrupting potential of phthalates. In: Metzler M (ed) Endocrine disruptors, Part I, vol 3, The handbook of environmental chemistry, Part L. Springer, Heidelberg, Germany, pp 169–201

Hilal SH (2006) Estimation of hydrolysis rate constants of carboxylic acid ester and phosphate ester compounds in aqueous systems from molecular structure by SPARC. U.S. Environmental Protection Agency. EPA/600/R-06/105

Hilal SH, Karickhoff SW, Carreira LA (2003) Prediction of chemical reactivity parameters and physical properties of organic compounds from molecular structure using SPARC. EPA/600/R-03/030

Hoffmann MR, Martin ST, Choi W, Bahnemann DW (1995) Environmental applications of semiconductor photocatalysis. Chem Rev 95:69–96

Huang PC, Kuo PL, Chou YY, Lin SJ, Lee CC (2009) Association between prenatal exposure to phthalates and the health outcome of newborns. Environ Int 35(1):14–20

IHCP (2008) Bis (2-ethylhexyl) Phthalate (DEHP) Summary Risk Assessment Report. Institute for Health and Consumer Protection (IHCP). European Commission. European Communities, 2008. Available: http://cerhr.niehs.nih.gov/chemicals/dehp/DEHPMonograph.pdf. Accessed Aug 2012

Jianlong W, Lujun C, Hanchang S, Yi Q (2000) Microbial degradation of phthalic acid esters under anaerobic digestion of sludge. Chemosphere 41:1245–1248

Jonsson S, Ejlertsson J, Svensson BH (2003) Behaviour of mono- and diesters of o-phthalic acid in leachates released during digestion of municipal solid waste under landfill conditions. Adv Environ Res 7:429–440

Juneson C, Ward OP, Singh A (2001) Biodegradation of bis(2-ethylhexyl) phthalate in a soil slurry-sequencing batch reactor. Process Biochem 37:305–313

Kelly TJ, Mukund R, Spicer CW, Pollack AJ (1994) Concentrations and transformations of hazardous air pollutants. Environ Sci Technol 28(8):379A–387A

Kirby AJ (1972) Hydrolysis and formation of esters of organic acids. In: Bamford CH, Tipper CFH (eds) Comprehensive chemical kinetics, vol 10. Elsevier, Amsterdam, pp 57–202

Kleerebezem R, Pol LWH, Lettinga G (1999) Anaerobic biodegradability of phthalic acid isomers and related compounds. Biodegradation 10:63–73

Lertsirisopon R, Soda S, Sei K, Ike M (2009) Abiotic degradation of four phthalic acid esters in aqueous phase under natural sunlight irradiation. J Environ Sci 21:285–290

Liang D, Zhang T, Fang HHP, He J (2008) Phthalates biodegradation in the environment. Appl Microbiol Biotechnol 80:183–198

Liu SM, Chi WC (2003) Effects of the Headspace gas composition on anaerobic biotransformation of *o*-, *m*-, and *p*-toluic acid in sediment slurries. J Environ Sci Health A Tox Hazard Subst Environ Eng 38(6):1099–1113

Mabey W, Mill T (1978) Critical review of hydrolysis of organic compounds in water under environmental conditions. J Phys Chem 7(2):383–415

Mabey WR, Smith JH, Podoll RT, Jonson HL, Moll T, Chou TW, Gates J, Partridge IW, Vandenberg D (1982) Aquatic fate process data for organic priority pollutants. U.S. Environmental Protection Agency Report. EPA 440/4-81-014

Mackintosh CE, Maldonado JA, Ikonomou MG, Gobas FAPC (2006) Sorption of phthalate esters and PCBs in a marine ecosystem. Environ Sci Technol 40(11):3481–3488

Matsumoto M, Hirata-Koizumi M, Ema M (2008) Potential adverse effects of phthalic acid esters on human health: a review of recent studies on reproduction. Regul Toxicol Pharmacol 50:37–49

Mersiowsky I, Ejlertsson J, Stegmann R, Svensson BH (1999) Long-term behaviour of PVC products under soil-buried and landfill conditions. Report for Norsk Hydro ASA, ECVM, ECPI, ESPA and ORTEP, Hamburg, Germany

Mersiowsky I, Weller M, Ejlertsson J (2001) Fate of plasticised PVC products under landfill conditions: a laboratory-scale landfill simulation reactor study. Water Res 35(13):3063–3070

Miller FC (1992) Composting as a process based on the control of ecologically selective factors. In: Blaine-Metting F (ed) Soil microbial ecology: applications in agriculture environment management. Marcel Dekker, New York, p 646

Nagao T, Ohta R, Marumo H, Shindo T, Yoshimura S, Ono H (2000) Effect of butyl benzyl phthalate in Sprague-Dawley rats after gavage administration: a two-generation reproductive study. Reprod Toxicol 14:513–532

OECD (1981) Hydrolysis as a function of pH. OECD GUIDELINE FOR TESTING OF CHEMICALS. http://www.oecd.org/dataoecd. Accessed on 7 May 2012

Oh BS, Jung YJ, Oh YJ, Yoo YS, Kang JW (2006) Application of ozone, UV and ozone/UV processes to reduce diethyl phthalate and its estrogenic activity. Sci Total Environ 367:681–693

Patnaik P, Yang M, Powers E (2001) Kinetics of phthalate reactions with ammonium hydroxide in aqueous matrix. Water Res 35(6):1587–1591

Peterson DR, Staples CA (2003) Degradation of phthalate esters in the environment. In: Staples CA (ed) The handbook of environmental chemistry, Vol. 3, Part Q. Springer, New York, pp 85–124

Sayyed HS, Mazhar FN, Gaikwad DD (2010) Kinetic and mechanistic study of oxidation of ester by KMnO$_4$. Int J ChemTech Res 2(1):242–249

Schierow L, Lee MM (2008) Congressional Research Service Report RL34572: Phthalates in Plastics and Possible Human Health Effects. Available at www.policyarchive.org/handle/10207/bitstreams/19121.pdf. Accessed Aug 2012

Schwartzenbach RP, Gschwend PM, Imboden DM (1992) Environmental organic chemistry. Wiley, New York

Schwarzbauer J, Heim S, Brinker S, Littke R (2002) Occurrence and alteration of organic contaminants in seepage and leakage water from a waste deposit landfill. Water Res 36:2275–2287

Schwarzbauer J, Heim S, Krooss B, Littke R (2006) Analysis of undisturbed layers of a waste deposit landfill—insights into the transformation and transport of organic contaminants. Org Geochem 37:2026–2045

Sharpe RM, Shakkebaek NE (1993) Are estrogens involved in falling sperm counts and disorder of the male reproductive tract? Lancet 341:1392–1395

Shen OX, Du GZ, Sun H, Jiang Y, Wu W, Song L, Wang XR (2009) Comparison of in vitro hormone activities of selected phthalates using reporter gene assays. Toxicol Lett 191:9–14

Shibata K, Fukuwatari T, Sasak R (2007) Phthalate esters enhance quinolinate production by inhibiting amino-carboxymuconate-semialdehyde decarboxylase (ACMSD), a key enzyme of the tryptophan-niacin pathway. Int Congr Series 1304:184–194

Skrzypek J, Lachowska M, Kulawska M, Moroz H (2008) Synthesis Of Bis(2-Ethylhexyl) phthalate over methane sulfonic acid catalyst, kinetic investigations. React Kinet Catal Lett 93(2):281–286

Stanley MK, Robillard KA, Staples CA (2003) Introduction. In: Staples CA (ed) The handbook of environmental chemistry, Vol. 3, Part Q. Springer, Berlin, pp 1–7

Staples CA, Peterson DR, Parkerton TF, Adams WJ (1997) The environmental fate of phthalate esters: a literature review. Chemosphere 35(4):667–749

Staples CA, Parkerton TF, Peterson DR (2000) A risk assessment of selected phthalate esters in North American and Western European surface waters. Chemosphere 40:885–891

Strac IV (2009) Migration of Di-(2-ethylhexyl) phthalate in normal saline during one year. Toxicol Lett 189:S259–S259

Sykes P (1975) A guidebook to mechanism in organic chemistry, 4th edn. Longman Group Ltd, London, pp 232–239

US EPA (1996) Method 8061A—Phthalate esters by Gas Chromatography with Electron Capture detection (GC/ECD). Available: http://www.epa.gov/epawaste/hazard/testmethods/sw846/pdfs/8061a.pdf. Accessed Aug 2012

US EPA (2001) Removal of endocrine disrupter chemicals using water treatment processes. 625/R-00/015, Washington DC

Wensing M, Uhde E, Salthammer T (2005) Plastics additives in the indoor environment-flame retardants and plasticizers. Sci Total Envion 339:19–40

Wolfe NL, Steen WC, Bums LA (1980) Phthalate ester hydrolysis: linear free energy relationships. Chemosphere 9:403–408

Woodward KN (1988) Phthalate esters: toxicity and metabolism, vol I. CRC, Boca Raton, FL

Xu G, Li F, Wang Q (2008) Occurrence and degradation characteristics of dibutyl phthalate (DBP) and di-(2-ethylhexyl) phthalate (DEHP) in typical agricultural soils of China. Sci Tot Environ 393:333–340

Yan H, Ye C, Yin C (1995) Kinetics of phthalate ester biodegradation by *Chlorella pyrenoidosa*. Environ Toxicol Chem 6:931–938

Zeng F, Cui K, Xie Z, Wu L, Luo D, Chen L, Lin Y, Liu M, Sun G (2009) Distribution of phthalate esters in urban soils of subtropical city, Guangzhou, China. J Hazard Mater 164:1171–1178

Zhang W, Li Y, Wang C, Wang P (2011a) Kinetics of heterogeneous photocatalytic degradation of rhodamine B by TiO2-coated activated carbon: roles of TiO_2 content and light intensity. Desalination 266:40–45

Zhang Z, Hu Y, Zhao L, Li J, Bai H, Zhu D, Hu J (2011b) Estrogen agonist/antagonist properties of dibenzyl phthalate (DBzP) based on in vitro and in vivo assays. Toxicol Lett 207:7–11

Zheng Z, He PJ, Shao LM, Lee DJ (2007) Phthalic acid esters in dissolved fractions of landfill leachates. Water Res 41:4696–4702

The Environmental and Human Effects of Ptaquiloside-Induced Enzootic Bovine Hematuria: A Tumorous Disease of Cattle

Rinku Sharma, Tej K. Bhat, and Om P. Sharma

Contents

1   Introduction 54
2   Enzootic Localities and Incidence 54
3   Clinical Symptoms 55
4   Etiology 57
    4.1   Bracken and Other Ferns 57
    4.2   Bovine Papillomavirus 64
    4.3   Interaction of Bracken Fern and Bovine Papillomavirus 65
5   Pathology 66
6   Experimental Animal Models 67
7   Effects on Human Health 73
    7.1   Ptaquiloside in the Human Food Chain 73
    7.2   Effect of Ptaquiloside on Soil, and Surface, and Ground Water 75
8   Diagnosis 79
    8.1   Gene Mutations 79
    8.2   Immunohistochemical Expression of Tumor Biomarkers 80
    8.3   Chromosomal Aberrations 81
    8.4   Oxidative Stress 82
    8.5   Ultrasonography 82
    8.6   Polymerase Chain Reaction for Bovine Papillomavirus 83
    8.7   Real-Time PCR for BPV 84
    8.8   DNase-SISPA for BPV 85
9   Prevention and Control 85
10 Future Areas for Research 87
11 Summary 88
References 88

R. Sharma (✉)
Disease Investigation Laboratory, Indian Veterinary Research Institute,
Regional Station, Palampur 176 061, Himachal Pradesh, India
e-mail: rinkusharma99@gmail.com

T.K. Bhat • O.P. Sharma
Indian Veterinary Research Institute, Regional Station, Palampur 176 061,
Himachal Pradesh, India

D.M. Whitacre (ed.), *Reviews of Environmental Contamination and Toxicology*,
Reviews of Environmental Contamination and Toxicology 224,
DOI 10.1007/978-1-4614-5882-1_3, © Springer Science+Business Media New York 2013

1   Introduction

Enzootic bovine hematuria (EBH) is a disease of cattle that is characterized by the intermittent presence of blood in urine caused by malignant lesions in the urinary bladder. The disease causes anemia, progressive emaciation, and death (Jones and Hunt 1983; Hopkins 1986). This neoplastic disease is primarily caused by chronic ingestion of ferns, and affliction with the papillomavirus is often associated with it (Campo et al. 1992; Campo 1997; Santos et al. 1998). A decade ago, a comprehensive review of this disease was published by Dawra and Sharma (2001). In this review, we provide an update on the various aspects of EBH and include an assessment of the effect of consuming these disease-inducing ferns on both animal and human health, and the associated environmental implications.

2   Enzootic Localities and Incidence

Cattle may resort to eating the bracken fern during dry seasons when other vegetation is sparse. Information on the incidence of the disease is usually based upon the abattoir surveys or on cases reported for treatment in hospitals. The number of cases reported for treatment represents only a small fraction of the total caused by the disease. This is known to be the case because there is no permanent cure for it, and in enzootic areas, farmers do not often report the disease. Hence, the exact incidence of the disease in a population of cattle is quite difficult to assess, except on a very local basis (Dawra and Sharma 2001).

Generally, a high incidence of EBH occurs in areas where the concentration of PT in the bracken is high (Smith et al. 1994a; Pinto et al. 2004; Marrero et al. 2004). Smith et al. (1994a) described one area where EBH occurred, by explaining that bracken fern commonly contained a large amount of PT (300–7,000 µg/g of dried bracken), whereas, in another area where there was no disease, the concentration was low—less than 100 µg/g dried bracken.

In an epidemiological investigation performed in seven localities of an enzootic area in Maramures, Baia Mare (Romania), it was found that bracken fern was consumed by cows, especially in the dry season when alternative vegetation was sparse (Giurgiu et al. 2008). There was no predisposition for one sex to consume more of the fern than the other one, although the disease appeared more frequently in females during pregnancy and after birth when clinical manifestations were more severe. EBH appeared mainly in winter months (January and February), and its development was associated with other competing diseases (placental reservation, prior and after birth paraplegia, renal ache, and lessening). The prevalence of cattle with clinical EBH caused by *Pteridium aquilinum* was evaluated between 2007 and 2008 in Brazil (Aparecida da Silva et al. 2009). One hundred and eighty-one bovines were examined, and 39 of the animals were sick. Among the affected cattle, 22 (56.4%) were affected with EBH.

In a retrospective study of 586 tumors encountered in Brazilian cattle over a 45-year period, tumors of the alimentary tract were those that were the most frequently encountered (139 cases). The incidence of this tumor type was attributed to a high incidence of squamous cell carcinoma of the upper alimentary tract that was associated with chronic ingestion of bracken fern, *P. aquilinum*. Consuming this carcinogenic plant was also associated with a relatively high incidence (35 cases) of urinary bladder tumors (Lucena et al. 2011).

In India, despite the widespread distribution of the fern in the Himalayas and hills of the southern country, the disease is restricted to well-defined pockets. The possible reason for its absence in other similar areas, even in presence of *P. aquilinum*, could be a lack of carcinogenic potential in the plant growing in such areas. The carcinogenic potential of *P. aquilinum* collected from an EBH-free hilly area in India has been studied (Dawra et al. 2002). In this study, the fern was fed to guinea pigs for 30 months at a rate of 30% w/w. The fern caused toxic and neoplastic changes in urinary bladders of the exposed animals. The incidence of tumors was 18.6%. The fern tested positive for the carcinogenic ptaquiloside (PT) (3.74 ± 0.6 mg/kg). The low level of PT, as compared to values reported elsewhere (0–9,776 mg/kg; mean 1,257 mg/kg), and grazing during periods when grasses are abundant are perhaps the reasons for the absence of the disease in such areas.

3   Clinical Symptoms

Hopkins (1987) investigated EBH for the first time in East Nepal, where *P. aquilinum* was abundant. This author found that the affected animals exhibited a chronic, intermittent hematuria and an associated polyuria and dysuria. The average age of onset of clinical signs was 7.3 years, and the duration was 1.5 years. Dawra et al. (1991) studied the enzymatic profile of urine and plasma in field cases of bovine bladder cancer. Urinary lactate dehydrogenase activity was significantly altered, as was the isoenzyme pattern. The activity of alkaline phosphatase and beta-glucuronidase was decreased in the affected animals. No significant changes were observed in acid phosphatase or arylsulphatase A and B activity. In plasma, lactate dehydrogenase activity was elevated, without any change in the isoenzyme pattern. No significant changes were observed in the other plasma enzymes studied or in the sialic acid concentration (Table 1).

A chronic toxicity study of *P. aquilinum* in cattle in the humid Chaco of Tarija, Bolivia, revealed both EBH and esophageal carcinoma in the affected herds. Sick animals showed cachexia, anemia, leucopenia, and urine that turned from pink to intense red color, with the presence of blood clots. Urinary bladder carcinomas and esophageal carcinomas were present in 100 and 50% of the examined cattle, respectively (Marrero et al. 2001). Two livestock farms in Spain, where EBH was known to occur, were studied for the clinicopathological findings (Perez-Alenza et al. 2006). Anemia, leucopenia, monocytosis, thrombocytopenia, hypergammaglobulinemia, microhematuria, and proteinuria were detected in the affected cattle. Three phases of the disease showed increasing severity and were established from multivariate statistical analyses. An initial phase, characterized by an extremely high monocytosis and otherwise normal parameters, was observed; an intermediate

Table 1 Clinical symptoms, enzymatic profile, and blood biochemical analysis of Enzootic bovine hematuria (EBH)-affected cattle

Particulars	Clinical symptoms/enzymatic profile/ biochemical profile	References
EBH-affected cattle in East Nepal, where *P. aquilinum* is abundant, age of onset of disease—7.3 year and duration—1.5 year	Chronic, intermittent hematuria and an associated polyuria and dysuria	Hopkins (1987)
EBH-affected cattle in the enzootic areas of Himachal Pradesh, India	Decrease in urinary alkaline phosphatase and beta-glucuronidase activity. Increase in plasma levels of lactate dehydrogenase	Dawra et al. (1991)
EBH-affected cattle in the humid Chaco of Tarija, Bolivia, chronic toxicity of *P. aquilinum*	Cachexia, anemia, leucopenia, and urine that turned from pink to intense red color, with the presence of blood clots	Marrero et al. (2001)
EBH-affected cattle of livestock farm in Spain	Anemia, leucopenia, monocytosis, thrombocytopenia, hypergammaglobulinemia, microhematuria, and proteinuria	Perez-Alenza et al. (2006)
EBH-affected cattle in Romania	Anemic mucous membranes, dysuria, and hematuria. Urinary sediment comprised of vesicle cells, red blood cells, and calcium carbonate and sulfate crystals. Serum hypoglycemia, hypoproteinemia with hypogammaglobulinemia	Giurgiu et al. (2008)
EBH-affected cattle in Brazil	Macrohematuria and low hematocrit values, ranging between 8 and 31% (normal range 31–47%) and leukocytosis ($132 \times 10^3/\text{mm}^3$)	Aparecida da Silva et al. (2009)

phase, characterized by monocytosis and moderate changes in other analytes, was seen; and a final phase was seen, characterized by normal levels of monocytes and many changes in other analytes. Monocytosis, detected in 31% of the younger animals, could represent an initial response to consumption of bracken fern and may be useful as an early hematological marker of EBH (Table 1).

Clinical examination of EBH-affected cows in Romania revealed apparently anemic mucous membranes, dysuria, and hematuria (Giurgiu et al. 2008). Hematuria appeared in the months of January, February, and March and occurred frequently in pregnant animals and in animals of approximately 2–3 years of age. Urine color varied between yellow and straw yellow, orange and pale red, had a pH range of 2.0–8.0, and showed urinary sediment comprised of vesicle cells, red blood cells, and calcium carbonate and sulfate crystals. The blood biochemical analysis showed hypoglycemia, hypoproteinemia with hypogammaglobulinemia, without significant changes of blood urea nitrogen (BUN) and serum creatinine (Table 1).

In Brazil, EBH-affected cows revealed macrohematuria, and the values of the hematocrit were low, ranging between 8 and 31% (normal range 31–47%). The mean total leucocyte count was $132 \times 10^3/\text{mm}^3$, and leukocytosis was present.

The temperature and capillary refill time were normal. Four animals (18.2%) out of the 22 affected ones had pale mucosal surfaces (Aparecida da Silva et al. 2009) (Table 1).

4 Etiology

4.1 Bracken and Other Ferns

Bracken is an invasive fern that comprises several subspecies and varieties that is prevalent throughout the world. It is ubiquitous in most tropical, subtropical, and some temperate countries (Santos et al. 1998). This fern has been described as one of the most common plants on the planet. The most common variety of bracken fern, *P. aquilinum* var. *aquilinum* (Fig. 1a, b), typically grows on sandy acid soils. Several plants contain mutagenic substances (Basaran et al. 1996), but it is reported that only bracken fern (*P. aquilinum*) causes cancer naturally in animals (Prakash et al. 1996). This disease causes high economic losses in cattle around the world (Wosiacki et al. 2005).

PT is a major genotoxic carcinogen (Mori et al. 1985), which has been isolated and characterized from *P. aquilinum*, and is considered to be the cause of most fern-associated animal health problems. PT content has been found to vary from 0 to 9,776 mg/kg in the *P. aquilinum* samples collected from different regions of the world (Smith et al. 1994b). PT, a sesquiterpenoid glycoside (Fig. 2), induces clastogenesis in cell cultures and has mutagenic and carcinogenic activity. Under acidic conditions, PT, which is a procarcinogen, undergoes aromatization by elimination of the glucose to give pterosin B, whereas under alkaline conditions, it is activated to give a conjugated dienone (activated ptaquiloside-APT, which is the ultimate carcinogen) after liberation of the sugar group (Ojika et al. 1987). APT alkylates DNA via a reactive cyclopropyl ring to form several DNA adducts (Ojika et al. 1987, 1989). The main labile adducts occur at the N-3 of adenine, and to a minor extent at the N-7 of guanine (Smith et al. 1994c; Kushida et al. 1994) within 24 h of exposure (Prakash et al. 1996). At the cellular level, affected cells have the capacity to repair such lesions in a short period of time, but a few of these lesions lead to mutations in key genes, and cancer may thereby be initiated. Besides PT, other bracken constituents like quercetin, isoquercetin, ptesculentoside, caudatoside, astragalin, and tannins may be carcinogenic as well (Dawra and Sharma 2001).

The PT concentrations in different varieties of bracken fern vary. Following are amounts found in different fern varieties: *P. aquilinum* var. *aquilinum*, *P. aquilinum* var. *esculentum* (280–13,300 µg/g) (Rasmussen et al. 2008), *P. aquilinum* var. *caudatum* (1,980–3,900 µg/g), and *P. aquilinum* var. *arachnoideum* (32–660 µg/g) (Alonso-Amelot et al. 1995). There is a positive correlation between elevation and the PT content of bracken. The concentration of PT is reported to be 2–3 times higher at high elevations (viz., 1,800 m) vs. low ones (viz., 1,000 m) (Villalobos-Salazar et al. 1999). PT is detected in all tissues of the bracken fern, and high levels

Fig. 1 (a) Bracken fern with young frond. (b) Young Bracken fern plant

Fig. 2 Ptaquiloside

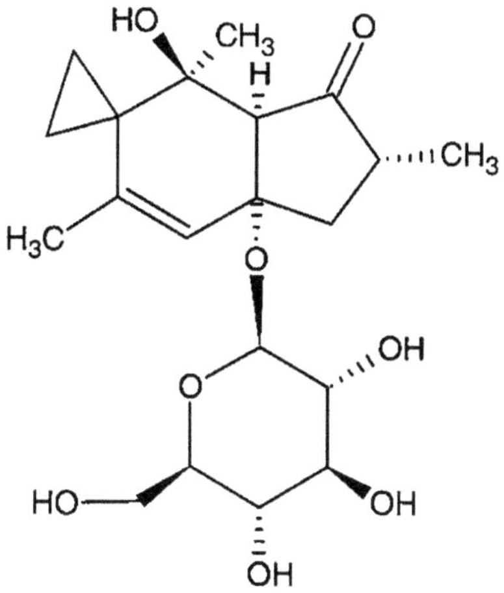

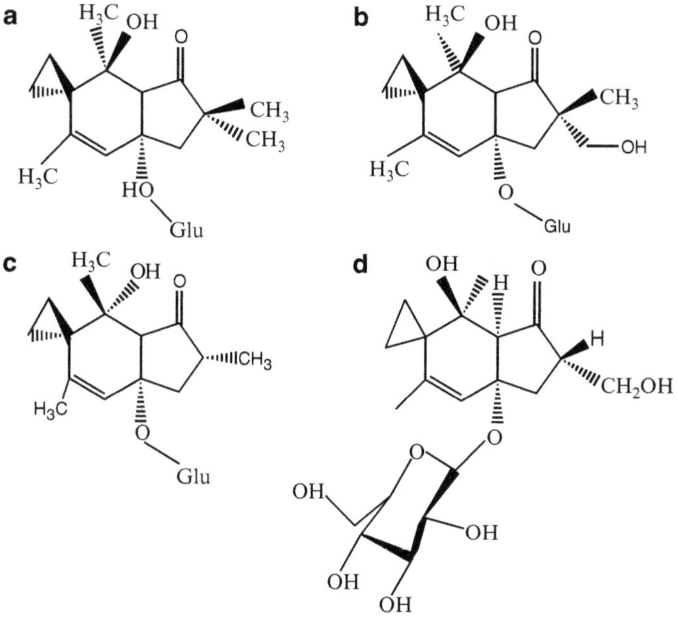

Fig. 3 Ptaquiloside Z (**a**), Caudatoside (**b**), Iso-ptaquiloside (**c**), Ptesculentoside (**d**) Reference: Rasmussen (2003); Fletcher et al. (2011)

are found in the crosiers (Rasmussen and Hansen 2003). PT is found in ferns other than bracken, namely, *Cheilanthes farinosa* (Forsk.) Kaulf., *Cheilanthes sieberi* Kunze, *Dryopteris juxtaposita* Christ., *Histiopteris incisa* (Thunb.) J. Sm., *Pteris cretica* L., and *Onychium contiguum* Hope (Smith et al. 1989; Saito et al. 1989, 1990; Agnew and Lauren 1991; Gounalan et al. 1999; Kumar et al. 2001). It is also reported that PT-like substances, such as ptaquiloside Z, caudatoside, and isopta-quiloside (Fig. 3a, b, c), may also cause EBH (Castillo et al. 2000). Fletcher et al. (2011)

determined the concentration of three unstable norsesquiterpene glycosides: ptaquiloside, ptesculentoside (Fig. 3d), and caudatoside and their respective degradation products, pterosin B, pterosin G and pterosin A, by HPLC (high pressure liquid chromatography) analysis in *Pteridium esculentum* and *P. aquilinum*. Samples of *P. esculentum* collected from six sites in eastern Australia contained up to 17 mg of total glycoside/gram dry wt. Both PT and ptesculentoside were present as major components and were accompanied by smaller amounts of caudatoside. Ratios of PT to ptesculentoside varied from 1:3 to 4:3, but in all Australian samples ptesculentoside was a significant component. This profile differed substantially from that of *P. esculentum* from New Zealand, which contained only small amounts of both ptesculentoside and caudatoside, with PT as the dominant component. A similar profile of PT as the dominant glycoside was obtained for *P. aquilinum* subsp. *wightianum* (previously *Pteridium revolutum*) from northern Queensland and also *P. aquilinum* from European sources. Ptesculentoside has chemical reactivity similar to that of PT, and presumably biological activity is also similar to PT (Fletcher et al. 2011). The presence of this additional reactive glycoside in Australian *P. esculentum* implies greater toxicity for consuming animals than previously estimated from PT content alone.

PT concentration has also been determined in 40 non-bracken fern samples collected, primarily from the northern mountainous state of Uttarakhand, India (Somvanshi et al. 2006). Of these, only *O. contiguum* contained high levels (viz., 499 and 595 mg/kg) of PT on a dry matter basis in the two samples collected. A few samples of *Diplazium esculentum, Polystichum squarrosum,* and *D. juxtaposita* showed moderate levels (19 to 31 mg/kg), one sample of *Christella dentata* had a very low level (0.4 mg/kg), but most samples lacked detectable PT. Samples of *O. contiguum* were collected from high-altitude areas of the Himalayas (Chamoli and Uttarkashi districts of Uttarakhand), where EBH is common. This fern was reported to induce ileal, urinary bladder, and mammary tumors on experimental feeding to guinea pigs (Dawra et al. 2001). In contrast, experimental feeding of *O. contiguum* to rats did not induce mortality or malignancy (Rajasekaran and Somvanshi 2001; Rajasekaran et al. 2004). The results of this study indicated that non-bracken fern species may also contain high levels of PT, and thereby may induce hazardous effects in animals, either alone or in combination with bracken fern (Somvanshi et al. 2006).

Saito et al. (1989) reported the distribution of PT and PT-like compounds in Pteridaceae as measured by chemical assay (thin layer chromatography; TLC) and the modified Ames test. They observed the widespread occurrence of such compounds in a variety of ferns, including *Cheilanthes myriophylla, Cibotium barometz, Dennstaedtia scabra, H. incisa, Pityrogramma calomelanos, P. cretica, Pteris nipponica, Pteris oshimensis, Pteris tremula,* and *Pteris wallichiana.* Smith et al. (1989) analyzed PT in *C. sieberi* samples collected from New Zealand and Australia using reversed-phase HPLC and indicated that this fern contained PT and other potentially carcinogenic pterosin B precursors. Agnew and Lauren (1991) also reported the presence of PT and related substances in the rock fern, *C. sieberi.*

Kumar et al. (2001) characterized the toxin from the fern *C. farinosa* and its effect on lymphocyte proliferation and DNA fragmentation. The aqueous extract of whole *C. farinosa* indicated the presence of PT or PT-like compound, with R_f values comparable to that of Pterosin B. HPLC analysis revealed the presence of 26.3 mg/kg PT. In vitro studies of the aqueous extract on lymphocyte culture revealed a correlation between stimulative indices and concentration of aqueous extract. Stimulation in lymphocyte proliferation was in the order of bracken > cheilanthes > ConA > PT.

Rasmussen et al. (2008) sampled *P. aquilinum* from 275 sites in New Zealand for estimation of PT levels. The fern was collected from both EBH-enzootic as well as non-enzootic areas. Three surveys were carried out for the study. In a regional survey, a total of 62 bracken stands were sampled from King Country, Waikato, and Coromandel regions of the North Island. In a farm survey, a sheep and cattle farm was sampled from the King Country region that had a history of EBH, and where high levels of PT had been encountered in earlier studies (Smith et al. 1988, 1992). A total of 26 stands were sampled in the farm survey. A national survey was also performed and included sampling nine areas from the North and South Islands, where EBH had previously been reported. These areas were identified from previous studies (Kerrigan 1926; Smith and Beatson 1970), from animal health laboratory records, and/or from communications from veterinary practitioners or meat veterinarians, and could only be defined broadly rather than confirmed to specific problem farms. A total of 186 stands of bracken were sampled from different ecosystems, including roadsides, fence lines, forests, bush, and grazed paddocks. The samples contained widely varying concentrations of PT (63% were positive), ranging from 280 to 13,300 (mean 3,800) μg/g (on a dry-wt basis). A high proportion of samples from the regional and national surveys, covering large areas of the country, contained no detectable levels of PT. The majority (61%) of samples from these two surveys that contained PT were collected from areas where EBH was present. Of the total samples, 42% contained PT from EBH endemic areas, compared with 6% from non-enzootic areas.

Recently, Nagarajan et al. (2011) detected PT levels in certain ferns of Nainital district, Uttarakhand state, and Nilgiris district, Tamil Nadu state in India. PT was detected in samples of five species: *P. cretica* (1,382.9 μg/g), *Asplenium dalhousiae* (807.9 μg/g), *Adiantum venustum* (756.1 μg/g), *Woodwardia unigemmata* (586.5 μg/g), and *Deparia japonica* (281.8 μg/g) collected from Uttarakhand. Similarly, the fern species *P. revolutum* (range 270.2–2,340.3, mean 1,189.0 μg/g), *Adiantum poiretii* (2,320.1 μg/g), *Doryopteris concolor* (388.1 μg/g), and *Pteris confuse* (362.3 μg/g), collected from Tamil Nadu, were also found to contain high to moderate levels of PT. The mean value for PT content in different fern samples collected was higher (661.83 μg/g) in the district of Nilgiris than in Nainital (240.86 μg/g). For the first time, PT was detected at very high to moderate levels in ten fern species *A. dalhousiae* (807.9 μg/g), *A. venustum* (756.1 μg/g), *W. unigemmata* (586.5 μg/g), *Athyrium mackinoniorum* (161.0 μg/g), *Onchyium lucidum* (51.8 μg/g), *Cheilanthes dalhousiae* (17.3 μg/g) from Nainital, Uttarakhand, and *A. poiretii* (2,320.1 μg/g), *D. concolor* (388.1 μg/g), *P. confuse* (362.3 μg/g), and

Pseudocyclosorus octhodes (63.3 µg/g) from Nilgiris, Tamil Nadu. The non-bracken ferns from Tamil Nadu state contained higher PT content than did samples collected in Uttarakhand state.

Bracken fern is a genotoxic carcinogen that exhibits organ specificity. It induces esophageal, intestinal, and bladder tumors in cattle and rats. APT is considered to be the ultimate carcinogen of bracken, with PT being its procarcinogen (Shahin et al. 1999). It has been proposed that the activation of PT into a mutagen under alkaline conditions may explain its organ specificity, since the pH in these organs is relatively high compared to others (van der Hoeven et al. 1983). Tumors of the esophagus, associated with a pH >8, specifically occur in cattle. In the rat, bracken fern predominantly induces tumors of the terminal section of the ileum, which has the highest pH in the intestine. PT may induce bladder cancer in cattle from the alkalinity that prevails in herbivorous animals. Since human urine is generally below pH 7, it has been proposed that bladder cancer from consuming bracken may be unlikely (van der Hoeven et al. 1983). Under alkaline conditions, PT undergoes a transformation involving the splitting off of the glucose moiety from the molecule to form ptaquilosin, and then a further splitting off of a hydroxyl group to form an illudane-dienone compound (APT) (Niwa et al. 1983; van der Hoeven et al. 1983; Shahin et al. 1999). APT has a greater capacity to alkylate DNA than PT because of its electrophilicity. PT has the ability to cross both lipid and hydrophilic barriers, because its aglycone conjugate is soluble in chloroform, methylene chloride, acetonitrile, and methanol, whereas the glucose fragment is water-soluble. It is thus conceivable that PT may penetrate deep into tissues and cells and reach endoplasmic and nuclear DNA, and thereby promote alkylation there (Alonso-Amelot and Avendano 2002). Although APT is stable in mildly alkaline conditions, it is immediately converted to pterosin B under weakly acidic conditions (Matsuoka et al. 1989).

It is known that alkylation of DNA bases at specific codons by certain chemicals may initiate carcinogenesis (Barbacid 1986). Much attention has been devoted to the chemical modification of proto-oncogenes. It is known that specific carcinogens such as nitrosomethylurea and dimethylbenzanthracene activate *H-ras* oncogenes by alkylating codons 12 or 61 (Jones et al. 1991). In the case of cattle under papillomavirus stress and eating bracken fern over a long period of time, activation of *H-ras* gene occurs in the upper alimentary canal. This activation ultimately produces esophageal carcinomas (Campo 1997; Campo et al. 1990) in the cattle and urinary bladder malignancies in rats (Masui et al. 1991). *H-ras* gene encodes for a GTP-binding protein, p21 and mutations interfere with its ability to regulate cell proliferation. Reddy and Randerath (1987) fed bracken fern to calves and then analyzed tissues and blood to detect if DNA-PT alkylation adducts were present. After being fed the bracken fern, platelets and neutrophils were depressed, which was regarded to indicate that the calves had been acutely poisoned by the bracken fern they consumed. The P1-nuclease-enhanced ^{31}P-postlabelling assay revealed that PT-DNA alkylation occurred only at specific organs, i.e., the ileum and urinary bladder, precisely those organs that are more prone to bracken-induced carcinogenesis in bovines (Reddy and Randerath 1987). Curiously, the hematopoietic system, so deeply affected otherwise, did not show the presence of PT-DNA adducts.

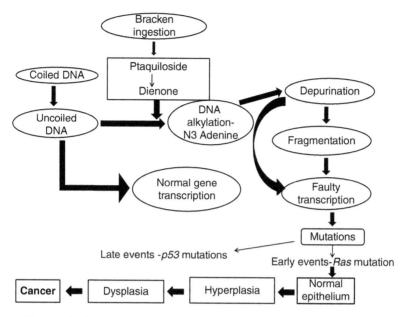

Fig. 4 The mechanism by which ptaquiloside induces carcinogenesis (Alonso-Amelot and Avendano 2002)

The ileum and bladder are known to be alkaline and hence provide ideal conditions for the transformation of PT to a strongly alkylating dienone (Fenwick 1988).

In addition, the *H-ras* gene showed an adenine to pyrimidine mutation on codon 61 that occurred quite rapidly, since adenine-PT alkylation at N3 occurred in 1 h only as observed in in vitro experiments, although guanines were also alkylated at N7 but at a much reduced pace (Reddy and Randerath 1987). Depurination and DNA fragmentation, a consequence of PT action that DNA synthases find difficult to repair, reached completion after only 22 h of incubation. It was inferred that PT induction of carcinogenesis in the ruminant occurs by organ-specific conditions (alkaline) for dienone formation, followed by early H-*ras* mutations (Shahin et al. 1998a), and that the true carcinogen is pteridine rather than PT (Shahin et al. 1998b). Shahin et al. (1999) hypothesized that mutations may take place at other key proto-oncogenes. Searches for PT-DNA adducts in the esophagus and stomach of humans who consumed bracken ferns have been made, but such adducts have not yet been found (Alonso-Amelot and Avendano 2002). In Fig. 4, we depict, in summary form, the mechanism by which ptaquiloside induces carcinogenesis.

Relevant studies have been performed in laboratory animals too. Freitas et al. (2001) reported that bracken-induced DNA adducts in mouse tissues are different from the adducts induced by the APT dienone. When DNA from esophagus, stomach, and ileum tissues was treated in vitro with dienone and analyzed by butanol extraction or nuclease P1 treatment, only one adduct was detected by ^{32}P-postlabeling. This adduct was not present in the DNA from mice treated with bracken fern or spores, suggesting that either bracken contains genotoxins other than PT or that the metabolism of PT produces genotoxins that are not reflected by APT. However, because the ATP-derived adduct has been detected previously in ileal DNA of bracken-fed calves, species-specific differences in the metabolism of bracken genotoxins may exist, thereby leading to differences in their biological outcomes. The signs of bracken poisoning in other animal species such as horses, mules, pigs, and rats have already been described by Dawra and Sharma (2001).

4.2 Bovine Papillomavirus

The bovine papillomaviruses (BPVs) are species-specific, double-stranded DNA viruses and produce cutaneous and mucosal neoplastic lesions. They are small non-enveloped viruses that possess an icosahedral capsid. Their open reading frames (ORFs) are divided into early (E) and late (L) regions. The early region encodes the nonstructural proteins E1 to E7, of which, E5, E6, and E7 are known to be oncoproteins. The late region encodes structural proteins L1 and L2, forming the capsid (Brandt et al. 2008; Chambers et al. 2003; de Villiers et al. 2004). BPV-2 may persist and may be maintained in a replicative status in the bloodstream, particularly in the lymphocytes. There, it may act as a reservoir of viral infection that in the presence of biological and/or chemical cocarcinogens could induce or influence bladder tumor development (Roperto et al. 2008). Evidence indicates that the malignant progression of urinary bladder lesions is dependent on an interrelationship between the BPV-2 infection and carcinogenic, mutagenic, and immunosuppressive compounds that exist in the bracken fern (Wosiacki et al. 2006).

Immunosuppression by BPV was sufficient to produce premalignant lesions, but it was the mutagens present in bracken that were responsible for their progression to neoplasia (Borzacchiello et al. 2003b). Upon cancer development, BPV-2 appeared to undergo significant changes, expressing the viral oncoprotein E5 and modifying telomerase activity (Borzacchiello et al. 2003b). A similar synergism between papillomavirus and bracken fern was also postulated to occur in bovine and human gastrointestinal tumors, associating BPV-4 and quercetin, a mutagenic flavonoid present in bracken fern (Carvalho et al. 2006).

In an attempt to better understand the biological role of circulating BPV-2 in EBH, an investigation was performed in which the simultaneous presence of BPV-2 in whole blood and urinary bladder tumors of adult cattle was evaluated (Roperto et al. 2008). Peripheral blood samples from 78 cattle, clinically suffering from EBH, revealed circulating BPV-2 DNA in 61 of them and also in two blood samples from healthy cows. Fifty of the affected animals were slaughtered at public slaughterhouses, and neoplastic proliferations in the urinary bladder were detected in all of them. BPV-2 DNA was amplified and sequenced in 78% of urinary bladder tumor samples and in 38.9% of normal samples as a control. Circulating episomal BPV-2 DNA was detected in 78.2% of the blood samples. The simultaneous presence of BPV-2 DNA in neoplastic bladder and blood samples was detected in 37 animals.

Specific viral E5 mRNA and E5 oncoprotein were also detected in blood by RT-PCR and western blot/immunocytochemistry, respectively. It is probable that BPV-2 persists and is maintained in an active status in the bloodstream, particularly in the lymphocytes, as a reservoir of viral infection; then, in the presence of cocarcinogens, this entity causes urinary bladder tumors to develop (Roperto et al. 2008).

BPV-2 DNA and the expression of E5 and E7 oncoproteins in vascular tumors of the urinary bladder in cattle were studied by Borzacchiello et al. (2007). EBH-affected cattle developed urinary bladder tumors of both epithelial and mesenchymal origin, mainly, hemangioma and hemangiosarcoma. The role of BPV-2, and of its major transforming oncoprotein in naturally occurring urothelial carcinogenesis, was clarified by Borzacchiello et al. (2003a). E5 interacted in vivo as in in vitro with the β receptor for the platelet-derived growth factor (PDGF). They showed that BPV-2 is present in 100% of the vascular tumors of the urinary bladder examined. Twenty-six out of 27 tumor samples (96%) expressed E5, whereas 20 out of 27 (74%) tumor samples expressed E7. The two viral oncoproteins were not expressed in normal endothelial cells. Additionally, they co-localized in neoplastic endothelial cells as demonstrated by confocal immunofluorescence. The PDGFβ receptor was also shown to be expressed and was co-localized with E5 in neoplastic blood vessels. These results demonstrated that the BPV-2 is present in high percentage in tumors of mesenchymal origin.

4.3   Interaction of Bracken Fern and Bovine Papillomavirus

Bracken fern has been reported to be carcinogenic when ingested by cattle or rats (Campo et al. 1992; Santos et al. 1987, 1992). However, researcher opinions on this are not unanimous, and reports of negative findings may be found. It has been documented that BPV-4 is one of the etiologic agents in alimentary tract carcinoma (Pennie and Campo 1992), and BPV-2 DNA is found in 46% of natural bladder cancers and in 69% of experimental bladder lesions (Campo et al. 1992) of animals raised on bracken fern.

A strong relationship between BPV-2 and bracken fern in both experimental and naturally occurring bovine bladder cancer was demonstrated by Campo and coworkers in 1992 (Campo et al. 1992). Although the synergism between the virus and bracken is still poorly understood, it is likely that BPV-2 infects the bladder mucosa to produce an abortive latent infection, as no structural proteins or virions are found in the urinary bladder mucosa (Borzacchiello et al. 2003b; Campo 2006). Chemical carcinogens from bracken may cooperate with BPV-2 to induce neoplastic disease, and it has been suggested that latent BPV is activated by bracken-induced immunosuppression, thus initiating progression to malignancy (Campo 1997).

Borzacchiello et al. (2003b) reported the association of BPV-4 and esophageal papillomas of cattle feeding on bracken fern in southern Italy. Esophageal lesions, i.e., single or multiple pedunculated proliferations or mucosal thickening were observed in 147 (13%) of the 1,133 slaughter house cattle examined, aged 4–12 years. BPV-4 was detected by PCR in more than 60% of the samples in which esophageal papilloma was diagnosed histopathologically.

Few studies have been carried out on the interaction of BPV and bracken fern in laboratory animals. Leishangthem et al. (2008) studied the pathological effects of BPV-2 and fern (*P. aquilinum* (L) Kuhn and *O. contiguum*) interaction in hamsters to evaluate early pathological changes. For tumor transmission studies in hamsters, 10% crude extract of cutaneous warts and urinary bladder tumor, were applied by scarification on abdominal skin of hamsters after 15 days of fern feeding. Cutaneous warts were successfully transmitted in hamsters approximately 3 months after inoculation, whereas urinary bladder tumors of EBH cases were not transmitted, even after 4 months (experimental trial duration). *O. contiguum* produced more pronounced effects than did bracken fern, which effects were characterized by a significant reduction in body weight and testicular atrophy. The authors suggested that long-term studies were required to induce urinary bladder tumors in hamsters, by feeding ferns followed by scarification of a rude extract of bladder tumors onto abdominal skin of hamsters.

Balcos et al. (2008) collected 90 urinary bladders from slaughtered cows from the hill-mountain area of Neamt County, Romania, wherein EBH was endemic. BPV-2 DNA was detected by PCR analysis in 68% of the analyzed tumor samples. BPV-2 positive urinary bladder tumors immunohistochemically expressed the major viral oncoprotein E5 intracytoplasmically and possessed a typical juxtanuclear pattern. E5 expression was not observed in normal mucosa, suggesting a causal role for this protein in the neoplastic process.

Resendes et al. (2011) assessed the presence of BPV-2 by PCR and immunohistochemistry for BPV E5 oncoprotein in urinary bladder lesions in cattle with EBH from the Azores archipelago, Portugal. An incidence rate of 28% of BPV-2 DNA in different types of tumors and cystitis cases (13 out of 46 samples) was recorded. The viral antigen immunolabeling was mainly detected within the cytoplasm of urothelial cells, displaying a juxtanuclear distribution.

5 Pathology

In a review of the bovine urothelial tumors and tumor-like lesions of the urinary bladder, Roperto et al. (2010) suggested the creation of a bovine bladder tumor classification system based on the 2004 WHO scheme (Epstein et al. 2004; Lopez-Beltran and Montironi 2004). Bovine urothelial tumors usually occur in multiples. Four distinct growth patterns of bovine urothelial tumors and tumor-like lesions are recognized: flat, exophytic or papillary, endophytic, and invasive. Carcinoma in situ (CIS) is the most common flat urothelial lesion, accounting for approximately 4% of urothelial tumors. CIS is detected adjacent to papillary and invasive tumors in 80–90% of bovine urothelial tumor cases. Approximately 3% of papillary lesions are papillomas, and approximately 5% are "papillary urothelial neoplasms of low malignant potential." Low-grade carcinoma is the most common urothelial tumor of cattle. High-grade carcinomas, and low- and high-grade invasive tumors, are less commonly seen.

Cattle naturally affected with EBH in Uttarakhand, India, revealed reduced hemogram, with low values of total erythrocyte count, hemoglobin, packed cell volume, and total leucocyte count (Singh 2007). On necropsy, three small papillae in rumen and multiple ecchymotic hemorrhages and papillae were seen in mucosa of urinary bladder in an EBH-affected cow. Histopathological evaluation revealed focal hyperplasia of rumen mucosa, and the urinary bladder tumor was diagnosed as transitional cell adenocarcinoma. Cytogenetic analysis revealed no significant changes in chromosome pattern of EBH-affected animal as compared to control (Singh 2007).

Carvalho et al. (2006) collected 433 urinary bladders with macroscopical lesions from the slaughterhouse of Sao Miguel Island (Azores, Portugal), an endemic area, wherein *P. aquilinum* infestation in pastures is high. Bladder lesions were divided into three main categories (viz., inflammatory, nonneoplastic epithelial abnormalities, and tumors). In some cases, neoplastic growth was localized to a single site, but usually multiple tumors developed within the same bladder. Epithelial tumors alone were present in 51.2% of the affected bladders, mesenchymal tumors alone in 17.4%, and both epithelial and mesenchymal tumors in the remaining 31.4%. The large number of tumors examined (870) revealed new categories not yet included in other veterinary classification systems, namely, inverted papilloma, papillary neoplasm of apparent low malignant potential, and hemangioendothelioma (Carvalho et al. 2006).

Brun et al. (2008) reported lymphoepithelioma-like carcinoma of the urinary bladder in a 7-year-old cow that had grazed pasture rich in bracken fern and had suffered from severe intermittent hematuria from 3 to 4 years of age. On necropsy examination there were multiple hemorrhagic foci scattered over the mucosal surface of the urinary bladder. Microscopically there were nests, cords, and sheets of neoplastic cells infiltrating the lamina propria and muscularis propria. These had a syncytial appearance with ill-defined cytoplasmic borders, large nuclei, and prominent nucleoli. There was a prominent associated inflammatory infiltrate comprising lymphocytes and plasma cells with sparse histiocytes and granulocytes. BPV-2 DNA was successfully amplified from the frozen neoplastic tissue and from selected areas of formalin-fixed, paraffin wax-embedded tissue obtained by laser capture microdissection.

6   Experimental Animal Models

Among the experimental animals studied, guinea pig has been the most used model for investigating fern-induced hematuria, urinary bladder cancer, and fern-induced poisoning, similar to that observed in cattle (Ushijima et al. 1983; Yoshida and Saito 1994; Dawra and Sharma 2001). In Table 2, we summarize the work performed in experimental animals by several workers who have conducted fern-feeding studies.

Table 2 A summary of the effects produced by feeding bracken and other ferns to laboratory animals

Laboratory animal	Feed	Course of experiment	Outcome	Remarks	References
Rats	Dried *P. aquilinum* 5% w/w	Exposure for 70 weeks	Ileal adenocarcinoma and sarcoma	Prolonged exposure, only ileal tumors	Santos et al. (1987)
Female rats	Dried, ground *P. esculentum* 25% w/w	Exposure for 162 days, observation for 15 weeks	Tumors in ileum and urinary bladder	Prolonged exposure, ileal and urinary bladder tumors	Smith et al. (1988)
Female rats and offspring	*P. aquilinum* 30% w/w	–	Reduced female fertility and weight gain during pregnancy, altered physical and neurobehavioral development in offspring	Adverse effect on female fertility and offspring	Gerenutti et al. (1992)
Guinea pigs	*P. aquilinum* 25% w/w, 30% w/w	Exposure for 100, 150 days	Urinary bladder tumors	Bracken fern induced bladder tumors in guinea pigs on short exposure	Bringuier et al. (1995)
Rabbits	*P. aquilinum* 25% w/w; *Dryopteris juxtaposita* 25% w/w	–; –	Mild to moderate vascular changes in most of visceral organs; More severe degenerative and vascular changes	Exposure did not result in tumors development in rabbits	Gounalan et al. (1999)
Guinea pigs	*O. contiguum* 30% w/w	Exposure for 30 months	Ileal, urinary bladder and mammary gland tumors	Prolonged exposure; Ileal, urinary bladder and mammary gland tumors in guinea pigs	Dawra et al. (2001)
Rats	*Polystichum squarrosum* 30% w/w	Exposure for 6 months	Early preneoplastic lesions in urinary bladder	Short exposure, preneoplastic bladder lesions	Sivasankar and Somvanshi (2001)
Guinea pigs	*P. aquilinum* 30% w/w	Exposure for 30 months, observation for 24 months	Urinary bladder and ileal tumors	Bladder and ileal tumors on prolonged exposure	Dawra et al. (2002)

Rats	*Diplazium esculentum* 30% w/w	Exposure for 30 days, observation for 30 days	Nil	Short exposure of *Diplazium esculentum* is nontoxic, noncarcinogenic to rats and guinea pigs	Gangwar (2004)
Guinea pigs	*D. esculentum* 30% w/w	Exposure for 30 days, observation for 30 days	Nil		
Rats	*P. aquilinum*	Exposure for 18 months	Molecular, subcellular, and cellular alternations and preneoplastic lesions in urinary bladder	Prolonged exposure did not produce tumors in rats and hill cattle	Ravisankar (2004)
Hill cattle	*P. aquilinum*	Exposure for 60 months			
Rats	*P. aquilinum*	Exposure for 6 months, observation for 2 months	Lowered neural activities	Neurotoxic effects due to presence of thiaminase enzyme in bracken fern	Ravisankar et al. (2005)
CD-1 mice	Ptaquiloside 0.5 mg/week, intraperitoneally	Exposure for 15 weeks, observation for 15 weeks	Urinary bladder urothelial dysplasia	Short exposure, early bladder neoplastic lesions	Costa et al. (2011)
Wistar female rats, 60 days old	*Pteris aquilina*	Exposure for 1 month	Brain lesions observed	Brain lesions due to presence of thiaminase enzyme in *P. aquilina*	Farajzadeh et al. (2011)

Santos et al. (1987) studied the effect of feeding *P. aquilinum* from Ouro Preto (Minas Gerais, Brazil) in rats. Fifteen (6 female and 9 male) 45-day-old Wistar rats were fed a diet containing dried bracken fern (5%, w/w) for 70 weeks. All experimental animals showed gastrointestinal tract tumors, which were located mainly in the ileum. Most tumors were malignant (adenocarcinomas and sarcomas), although benign adenomas were also present. One animal developed a lymphoma and none showed vesicle tumors (Table 2).

Smith et al. (1988) studied the carcinogenicity of *P. esculentum* in laboratory rats in New Zealand. *P. esculentum* was collected from two sites. At site 1, EBH was non-endemic, and at site 2 it was known to occur. The fern was dried, ground, and incorporated (25% w/w) into a pelleted diet and was then fed to female rats for a period of 162 days. Fifteen weeks later, when the rats were autopsied, numerous tumors were found, mainly of the ileum and urinary bladder, and were present in the animals fed the bracken fern collected from site 2. Neoplasms were found in 85% of rats from the site 2 group and 11% from the site 1 group, whereas only a single tumor (a hemangioma of the uterus) was observed in the controls. Analysis of the fern and feed pellets (incorporated with fern) for PT showed much higher levels in the plant material from site 2. PT levels from site 1 were 26 µg PT/g of dried fern and 6.5 µg PT/g of feed pellets. Site 2 showed presence of 2,270 µg PT/g of dried fern and 355 µg PT/g of feed pellets (Table 2).

Gerenutti et al. (1992) studied the effect of feeding bracken fern (*P. aquilinum* L. Kuhn) on the development of female rats and on their offspring. When fed as 30% of the normal diet ad libitum, no effect was noted on the weight gain of the female rats during development. In addition, no effects at this dietary level were noted on the estrus cycle or on milk production. However, intake of the fern at the dose of 30% w/w, reduced female fertility and weight gain during pregnancy. It also adversely affected physical and neurobehavioral development of the offspring (Table 2).

Bringuier et al. (1995) studied bracken fern-induced bladder tumors in guinea pigs as a model for human neoplasia. The guinea pigs were fed a diet containing 25 or 30% dried bracken fern for 100 or 150 days. A high incidence of bladder tumors occurred. All but one animal had preneoplastic or neoplastic lesions after 4 months of exposure to the fern diet; after 1 year, 24 of 25 exposed animals had carcinomas. The bladder tumors that developed were essentially pure transitional cell carcinomas, although areas of focal squamous metaplasia were observed in four cases. Immunohistological detection of cytokeratins 10, 13, and 18 confirmed the transitional nature of these tumors. Dysplasia and preneoplastic hyperplasia were seen after 4 months of dietary exposure, and papillary carcinomas appeared after 6 months, whereas muscle-invasive carcinomas required 1 year of exposure. Thus, the full spectrum of preneoplastic and neoplastic bladder lesions observed in humans was reproduced in this model. Interestingly, when tumors were induced in older guinea pigs, none progressed to a muscle-invasive stage. This phenomenon should provide the opportunity to study the molecular mechanisms associated with these two different growth patterns, a major issue in understanding human as well as bovine bladder tumor progression (Table 2).

Gounalan et al. (1999) studied the effect of *P. aquilinum* and *D. juxtaposita* ferns in laboratory rabbits. The fern was included at a dose of 25% in concentrate ration mixture. The exposed rabbits were affected and displayed progressive anemia, leucopenia, lymphopenia, and relative heterophilia. Significant elevations in serum enzymes, like serum glutamate oxaloacetate transaminase, serum glutamate pyruvate transaminase, alkaline phosphatase, urea, and creatinine levels, were seen. Histopathologically, rabbits revealed mild to moderate vascular changes in most visceral organs, vacuolar degenerative changes in hepatocytes, hypersecretory activity in intestine, the presence of casts in renal tubules, and degenerative changes in renal tubular-lining epithelial cells. The authors observed that *D. juxtaposita*-fed rabbits showed more severe degenerative and vascular changes than those to *P. aquilinum*.

In a long-term *O. contiguum* fern-feeding trial of 30 months on guinea pigs in India, intestinal tumors were found in the ileal region of four animals, and preneoplastic lesions were observed in another four animals (Dawra et al. 2001). There was thickening of the urinary bladder wall, edema, and ulceration, desquamation of the epithelium, hemorrhages, nodular and papillary hypoplasia, and proliferation of the venules in the exposed animals. In one animal, the changes advanced to the metaplastic stage, and in another, they progressed to the neoplastic stage. Microhematuria appeared in five animals, but did not develop to macrohematuria by the end of the experiment. In one animal, mammary gland tumors developed (Dawra et al. 2001).

Sivasankar and Somvanshi (2001) observed moderate mortality, decreased body weight, less body fat, and splenomegaly in rats fed *P. squarrosum* (D. Don) at a dose of 30% w/w for 6 months. On postmortem examination, other than splenomegaly no significant gross lesions were seen in sacrificed animals. Histopathologically, *P. squarrosum*-fed rats showed dilated Virchow-Robin's space in brain, mild to moderate vascular changes like edema, engorgement of blood vessels and hemorrhages in most of the visceral organs, interstitial pneumonia in lungs, focal necrosis and generalized vacuolative degenerative changes in liver, more hemosiderin deposition, and presence of a higher number of megakaryocytes in spleen. The kidneys revealed shrunken glomeruli, increased periglomerular space, and more glomeruli per microscopic field. Focal hyperplasia of urinary bladder was noticed. Testes showed moderate to marked depletion of germinal epithelium and spermatids in seminiferous tubules. Pathologically, progressive changes were observed only in liver, urinary bladder, and testes on 180 days post feeding (DPF). One fern-fed rat sacrificed on 135 DPF showed a hepatic tumor that was diagnosed as hepatocellular carcinoma. The results showed that *P. squarrosum* produced early preneoplastic lesions as has been reported in bracken fern-fed animals (Sivasankar and Somvanshi 2001).

In an experimental study, bracken fern, *P. aquilinum* (collected from non-enzootic area of Himachal Pradesh, India) was fed to guinea pigs for 30 months at the rate of 30% (w/w). Results were that the fern diet caused toxic and neoplastic changes in urinary bladders of the exposed animals (Dawra et al. 2002). Grossly, urinary bladder congestion, edema, and hemorrhages to a varying extent were observed in exposed animals. Histopathologically, the animals ($n=3$) that died between 0 and

16 months of exposure did not show any change in the intestinal mucosa. But in these animals, mild to severe congestion, along with edema and hemorrhages, were observed in the lamina propria of urinary bladder. The urothelium revealed areas of desquamation and mild proliferation. With one exception, the animals ($n=5$) that died between 17 and 30 months did not show any changes in the intestines; one animal displayed proliferative changes in the caecum. Changes in urinary bladder were similar to those described for the animals that died between 0 and 16 months. The urinary bladder of one animal revealed an irreversible type of hyperplastic urothelium. Animals ($n=8$) that died, or were sacrificed between 31 and 54 months, showed changes to both intestine and urinary bladder (Dawra et al. 2002). In three animals, proliferative changes were observed in the epithelial mucosal lining of the intestine. One animal revealed proliferative changes in the ileal-, and cecal-mucosal epithelial lining. Squamous metaplasia of the ileal lining cells was observed in a third animal. Proliferation of goblet cells and congestion in lamina propria, along with inflammatory cell infiltration, were observed in most animals. Three animals revealed malignant transformation in the epithelial lining of the bladder mucosa; this was in the form of adenocarcinoma in two animals and transitional cell carcinoma in one animal. The animal with transitional cell carcinoma revealed severe proliferation of transitional epithelium in the form of sheets, and the growth was infiltrated in the lamina propria. The urothelium of control animals did not reveal any histological changes. During the exposure period, screening of urine samples showed microhematuria in one animal. Three more animals developed hematuria between 31 and 54 months (Dawra et al. 2002) (Table 2).

Gangwar (2004) did not observe any pathological effects of *D. esculentum*, Retz. in laboratory rats and guinea pigs, after being fed the fern in feed for 30 days at a dose of 30% w/w. The animals were observed for another 30 days after the fern was discontinued from the diet, and no effects were seen. Ravisankar (2004) studied the long-term pathological effects of bracken fern (*P. aquilinum*) in laboratory rats and in hill cattle. This author reported that feeding bracken fern to rats for 18 months caused clinicopathological changes similar to those observed in cows fed these ferns for 60 months. Prolonged feeding of bracken fern, in which low levels of PT existed, failed to induce neoplasia in either species. However, such feeding produced molecular, subcellular and cellular alternations, and preneoplastic lesions. Behavioral and pathological effects (Ravisankar et al. 2005) were observed in rats fed a crude extract of bracken fern for 6 months. After these animals were observed for an additional period of 2 months, they displayed lowered neural activities exhibited by dullness, lack of feeding, opisthotonus, tremors in hindlimbs, reduced number of ambulations, and occasional spastic paralysis of hindlimbs at various periods during the experiment. A significant decrease in body weight and an increase in brain weight were observed. Histopathological lesions involving engorged blood vessels were seen, along with thickened vessel walls, hemorrhages in meninges and stroma, and occasional glial cell proliferation in the brain; these and other clinical observations were suggestive of neurotoxic effects induced from dietary exposure to bracken fern. It was concluded that the neurotoxic effects of the crude-extract drenched groups of rats may have been caused by the presence of thiaminase enzyme in bracken fern (Table 2).

PT-induced, B-cell lymphoproliferative, and early-stage urothelial lesions in mice were recently studied by Costa et al. (2011). A total of 12 male CD-1 mice were intraperitoneally administered PT at a dose of 0.5 mg weekly for 15 weeks, and the animals were then observed for another 15 weeks after treatment ceased. The 12 animals used as controls were administered only the vehicle solution (phosphate buffered saline). All ten surviving mice developed a lymphoproliferative malignancy. Two mice died during the course of the experiment. The observed lymphoproliferative disease that was induced was characterized by multifocal B-(CD45þ/CD3)-lymphocytic renal (10/10 animals) and hepatic (2/10 animals) invasion, splenic white pulp hyperplasia (10/10), together with a significant increase in circulating B-(CD19þ)-lymphocytes and the appearance of circulating dysplastic lymphoid cells. Eight of ten PT-exposed animals developed urothelial dysplasia (six low-grade dysplasia and two high-grade dysplasia). No lesions were detected in control mice. These results showed that PT was capable of inducing malignant transformations in mice, and the study provided an in-depth characterization of the nature of the lymphoproliferative lesions that formed. The occurrence of urinary bladder preneoplastic lesions (low- and high-grade zonal urothelial dysplasia) in eight out of ten mice is a new finding in this species.

Farajzadeh et al. (2011) evaluated the chronic toxicity in rats of the fern *Pteris aquilina* that was harvested in northern Iran. Thirty native Wistar female rats, 60 days old, were segmented into groups that received either one of four treatments, and one was reserved as a control group. Groups 1–3 were, respectively, treated with fern at a dose of 5, 10, and 15%. The fourth animal group was fed 15% of their diet as *P. aquilina,* but the diet was admixed with vitamin B12 at a 2 mg/kg dietary dose rate. The control group of six animals served as a negative control and was fed normal rat chow. The experimental animals were sacrificed after 1 month when they demonstrated signs of poor health. After sacrifice, the animals were examined for gross lesions, particularly in the brain. The appearance of histologic changes in the brain was regarded to suggest that the experiment had produced lesions comparable to those that occurred naturally in bovine encephalomalacia. Such lesions included severe hemorrhages and vacuoles of various sizes in cerebrum, cerebellum, and brain stem. Necrosis (malacia) with gitter cell reaction was evident. The incidence of brain lesions was significantly greater in groups that received the 15% diet of *P. aquilina* without thiamine supplement. These results provided further evidence for the presence of an antithiamine substance in *P. aquilina* and led to the conclusion that *P. aquilina* itself is responsible for the induced rat-brain lesions (Table 2).

7 Effects on Human Health

7.1 *Ptaquiloside in the Human Food Chain*

An association between bracken consumption and cancer in humans has been previously demonstrated (Alonso-Amelot et al. 1998). The main routes of exposure include ingesting bracken crosiers, drinking PT-contaminated milk or water, and

Fig. 5 Ptaquiloside: points
of entry into the human food
chain

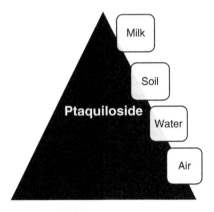

inhaling air that contains bracken spores (Fig. 5). Such exposures may lead to an increased incidence of gastric and esophageal cancer in humans. It has been found that approximately 9% of PT in bracken fern consumed by cows is transferred to the cow's milk (Hirono et al. 1972; Alonso-Amelot 1997; Marliere et al. 2000; Alonso-Amelot and Avendano 2002). Moreover, PT is transferred from bracken vegetation to the underlying soil, where it may leach to adjoining soil and ground water and thereby contaminate drinking water (Rasmussen et al. 2003a, b). The leaching of PT to the aqueous environment occurs most extensively in sandy soils that have a pH >4 and those that are low in organic matter content (Rasmussen et al. 2005).

Rasmussen et al. (2005) analyzed for PT concentrations in two wells that were located near infestations of bracken and respectively found levels of 30 and 45 mg/L in the water. This author also found PT levels up to 7 mg/L in the soil solution in samples taken 90 cm beneath a bracken stand. In a study by Siman et al. (2000), extracts from spores of different ferns, including *P. aquilinum*, were assessed for producing DNA damage. They found that *P. aquilinum* produced the maximum damage to the DNA. However, it is not yet known whether such damage will lead to tumor development. The concentration in air of spores over a sporulating population of bracken ferns may reach 800 spores/L. At this air concentration, the average human would inhale 50,000 spores during an exposure period of 10 min. In some countries, bracken is used as food and is sometimes ingested on a daily basis. Usually, bracken is boiled in water that contains wood ash or sodium bicarbonate before being consumed. Marliere et al. (2002) and Hirayama (1979) demonstrated that, for people in Brazil and Japan who consume bracken daily, the frequency of esophageal and gastric cancer is 5.5 and 8.1 times higher than nearby populations who do not eat bracken. In other studies, it has been revealed that people who live for more than 20 years (and are exposed during childhood) in areas that have the mere presence of bracken fern face a greater risk (3.6-fold increase) of death from gastric adenocarcinoma (Alonso-Amelot and Avendano 2002; Schmidt et al. 2005).

Fletcher et al. (2011) detected the residues of norsesquiterpene glycosides in the tissues of cattle fed Austral bracken (*P. esculentum*). Calves were fed a diet containing 19% *P. esculentum* that delivered 1.8 mg of PT and 4.0 mg of ptesculentoside/kg

Fig. 6 The flow of ptaquiloside to and through the human food chain

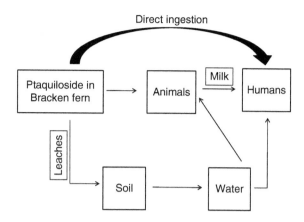

body wt/day to explore the carcass residue potential of these compounds. Concentrations of PT and ptesculentoside in the liver, kidney, skeletal muscle, heart, and blood of these calves were determined by HPLC-UV analysis of their respective elimination products, pterosin B and pterosin G. Plasma concentrations of up to 0.97 µg/mL PT and 1.30 µg/mL ptesculentoside were found but were shown to deplete to <10% of these values within 24 h of bracken consumption. Both glycosides were also detected in all tissues assayed, with ptesculentoside appearing to be more residual than PT. Up to 0.42 and 0.32 µg/g of ptesculentoside was present in skeletal muscle and liver, respectively, 15 days after bracken consumption ended. This level of residual glycosides in tissues of cattle feeding on Austral bracken raises health concerns for consumers and warrants further investigation. The pathway flow by which fern PT may enter the human food chain is depicted in Fig. 6.

7.2 Effect of Ptaquiloside on Soil, and Surface, and Ground Water

Rasmussen (2003) explored the possibility of PT being an environmental hazard by monitoring the occurrence of PT in terrestrial ecosystems and investigating PT stability and mobility in soils of Denmark. PT is leached from bracken field stands to the soil environment (Rasmussen et al. 2003b), where the compound is highly water-soluble and shows little sorption to soil particles (Rasmussen et al. 2005).

Rasmussen et al. (2005) originally reported that mildly acidic to neutral sandy soils would be the most prone to PT leaching. However, Ovesen et al. (2008) now report that clayey soils cannot be exempt from posing a risk because there is an apparent stabilizing effect of clay that retards PT degradation. Moreover, the presence of macropores in clay could substantially reduce residence times, and PT is also subject to leaching in clayey soils. Degradation is retarded by low temperatures and low organic matter content (Rasmussen et al. 2005), and it is more rapid for acid (pH <4) sandy soils, although this decreases with reduced organic matter content (Rasmussen et al. 2005). Organic matter plays an important role in binding of PT to the soil. Although PT has a relatively low sorption coefficient, the presence of organic matter reduces PT mobility and enhances degradation, which limits amounts that may reach water sources.

Rasmussen et al. (2003b) carried out a study to identify environmental parameters that correlate with the PT content in fronds, and to quantify the amount of PT in the soil environment. The content of PT in bracken fronds and topsoil materials was quantified in Danish Bracken stands to evaluate the possibility for predicting potential hot spots of potential PT leaching and to gain insight on factors causing production and leaching of PT. The content of PT in the fronds ranged from 110 to 3,800 μg/g of bracken, with a median content of 555 μg/g. The potential PT load related to the amount of PT found in the standing biomass at the end of the growing season. This is the amount of PT that will be transferred to the litter layer and subsequently become a part of the soil O-horizons, when the fronds die and degrade after the first October frost. The potential PT load was measured at different sites, and the levels found ranged between 10 and 260 mg/m^2 (frond density — number of fronds per m^2), with nine of the sites having potential loads exceeding 100 mg/m^2.

Jensen et al. (2008) developed a sensitive detection method to analyze for PT and its transformation product pterosin B in soil and ground water, by using liquid chromatography–tandem mass spectrometry (LC–MS/MS). Detection limits for the analyses were 0.19 μg/L (ptaquiloside) and 0.15 μg/L (pterosin B), which are 300–650 times more sensitive than for previously published LC-UV methods.

Rasmussen et al. (2003a) studied the distribution of PT in four Danish bracken populations to evaluate the degree to which PT is transferred from ferns to soil. Results were that populations showed statistically significant differences in PT content of their fronds and rhizomes. The highest concentrations were encountered in fronds, in which concentrations ranged from 213 to 2,145 μg/g; rhizomes had concentrations ranging between 11 and 902 μg/g. PT levels in soil were 0.22–8.49 μg/g but apparently bore no correlation with the PT concentration of fronds or rhizomes. Although the PT content in the Oi horizons (litter layer) could have been derived from the soil material itself, since most of the soil was composed of partly mineralized bracken fronds from previous years, this was unlikely, because PT is unstable and hydrolyzes at the low pH of the Oi horizons (Saito et al. 1989). Analysis showed that water could leach PT from bracken fronds, which result supports the high PT content in soil at sites exposed to heavy showers just before sampling. The observed soil content corresponded to estimated soil solution concentrations of 200–8,500 μg/L, demonstrating a substantial risk of PT contamination of surface and ground water. The high potential PT concentrations in pinnae drips and in soil water emphasize that soil and water contamination by PT is likely to occur and should be considered when investigating the potential link between human gastric cancer and bracken exposure.

Leaching of PT from soils is expected to increase in slightly acidic and neutral soils, where the degradation of PT occurs only slowly (Rasmussen et al. 2003a, b, 2005). In addition, low humic matter and clay content favor the mobility of PT

(Rasmussen et al. 2005). Schmidt et al. (2005) found that PT is toxic to soil microorganisms, which may lead to a slower biodegradation of the compound by these microbes. In this same study, a reduction of the actively respiring microbial population occurred as PT exposure increased in a Danish agricultural soil (calcareous loamy Mollisol). Subsequently, Engel et al. (2007) studied both the degradation and toxicity of PT in weakly acidic Danish and New Zealand soils that underlaid natural bracken stands. Non-sterile and sterile Danish soils from the A, E, B, and C horizons were spiked with pure PTA at a level of 25 μg/g. The specific aims were to evaluate the importance of microbial activity for degrading PT and to investigate the toxicity of PT to microbial communities in bracken-impacted soils. PT contents of 2.1 ± 0.5 mg/g and 37.0 ± 8.7 mg/g tissue were measured in bracken fronds from Denmark and New Zealand, respectively. PT levels in the two soils were similar (0–5 μg/g soil); a decrease with depth was discerned in the deeper B and C horizons of the Danish soil (weak acid sandy Spodosol), but not in the New Zealand soil (weak acid loamy Entisol). In the Danish soil, PT turnover was predominantly due to microbial degradation (biodegradation); chemical hydrolysis was occurring mainly in the uppermost A horizon, wherein the pH was very low (3.4). Microbial activity (basal respiration) and growth ([^{3}H] leucine incorporation assay) increased after PT exposure, indicating that the bracken toxin served as a carbon substrate for the organotrophic microorganisms. In contrast, PT apparently had no affect on microbe community size, when the measures used were substrate-induced respiration or composition. This demonstrated that PT stimulated microbial activity and that microorganisms play a predominant role in rapidly degrading PT in bracken-impacted soils (Engel et al. 2007).

PT leaches from plants, then to and through the soil into water channels (Rasmussen et al. 2003b). Therefore, it is necessary to assess PT toxicity in soil and water. Schmidt et al. (2005) performed a study in which they measured the inhibition of soil respiration and genotoxic activity of PT. To perform their work, the authors used a soil microbial carbon transformation test and an *umu* test. In the carbon transformation test, sandy loam soil was incubated at five different initial PT concentrations for a period of 28 days, after which glucose was added and respiration measured for 12 consecutive hours. The tests were performed at 20°C and at a soil moisture content of approximately 15%. For soil material sampled in the autumn, initial PT concentrations were tested that ranged from 0.008 to 40.6 mg PT/g dry soil. From fitting the resulting data to a sigmoidal function, a 10% effect dose (ED_{10}) was estimated to be 13 mg PT/g soil dwt, with an upper 95% confidence limit of 43 mg PT/g soil dwt, and a 95% lower confidence limit of 2 mg PT/g soil dwt. For soil material sampled in late winter, the initial PT concentrations tested ranged from 1.56 to 212 mg PT/g soil dwt; test results produced an ED_{10} value of 55 mg PT/g soil dwt, with an upper 95% confidence limit of 70 mg PT/g soil dwt, and a 95% lower confidence limit of 40 mg PT/g soil dwt.

The genotoxic activity of PT was determined by using the *umu* test, both without and with metabolic activation (addition of S9 rat liver homogenate, Commercial liver S9 fraction from Aroclor 1254) (Schmidt et al. 2005). In tests with the addition of S9, the induction ratio exceeded the critical ratio of 1.5 at a PT concentration of

46 ± 16 mg/mL, and, in tests without S9, the critical ratio was exceeded at a PT concentration of 279 ± 22 mg/mL. The genotoxicity of PT is comparable to that of quercetin, another bracken constituent. The toxicity of PT toward microorganisms prolongs the persistence of PT in terrestrial environments, increasing the risk of PT leaching to drainage and ground water. Both genotoxicity and inhibition of soil respiration by PT was observed. PT was genotoxic in concentrations of about 46 mg/mL at pH 7.4, without prior alkaline preactivation. Addition of S9 decreased the genotoxic activity of PT by at least a factor of 4, which indicates that PT can be detoxicated partly by the proteins and enzymes present in S9. Effects on microbial processes were evident in the two carbon transformation tests. Two independent estimates of ED_{10} of 13 and 55 mg PT/g soil dwt were made for the same soil, but sampled at two different times; results of the two estimates were not significantly different. These effects may lead to slow degradation of PT in soil environments. Poor degradation, together with high hydrophilicity and poor sorption of PT to soil, leaves a risk of PT leaching to the aqueous environment, thus causing contamination of surface and ground water.

The kinetics of PT degradation was subsequently studied in the aqueous environment by Ayala-Luis et al. (2006). The kinetics of PT hydrolysis was examined at 22°C in aqueous buffered solutions (pH 2.88–8.93). The resulting reaction followed first-order kinetics, with respect to PT under all pH and temperature conditions. At pHs lower than 4.43 (± 0.32), the reaction was acid-mediated, whereas the reaction was base-mediated at pHs higher than 6.39 (± 0.28). The rate constants for the acid-catalyzed, base-catalyzed, and neutral hydrolysis were 25.70 (± 0.96), 4.83 (± 0.03) $\times 10^4$, and 9.49 (± 6.02) $\times 10^4$/h, respectively. The PT hydrolysis at pH 4.46 is strongly dependent on temperature and has an activation energy of 74.4 (± 2.6) kJ/mol. Stoichiometric calculations, reaction kinetics, and ultraviolet–visible spectrophotometry strongly indicated the formation of an intermediary compound at pH 5.07 and 6.07, via a mechanism comprising two first-order consecutive reactions. PT has the lowest rate of hydrolysis at slightly acidic pH and low temperatures. Therefore, because PT is not materially sorbed onto soil, slightly acidic sandy soils in cold climates are most prone to the leaching of PT to deeper soil layers and aquifers.

Ovesen et al. (2008) quantified the degradation rates of PT in a laboratory at 10°C, in soil and soil solutions in sandy and clayey soils that were subjected to high natural PT loads from adjacent bracken stands. Degradation kinetics in moist soil could be fitted to indices of both the sum of a fast and a slow first-order reaction; the fast reaction contributed 20 to 50% to the total degradation of PT. The fast reaction was similar in all horizons, and the rate constant k (1F) ranged between 0.23 and 1.5/h. The slow degradation had the rate constant k (1S) and ranged between 0.00067 and 0.029/h. The latter was more than twice as fast in topsoils compared to subsoils, which is attributable to higher microbial activity in topsoils. Experiments with sterile controls confirmed that nonmicrobial degradation processes constituted more than 90% of the fast degradation and 50% of the slow degradation. The lower nonmicrobial degradation rate observed in the clayey compared with the sandy soil is attributed to a stabilizing effect of PT by clay silicates. PT appeared to be stable in all soil solutions, in which no degradation was observed within a period of 28 days.

This was in strong contrast to previous studies of hydrolysis rates performed in artificial aqueous electrolytes. In the present study, Ovesen et al. (2008) predicted that the risk of PT leaching is controlled mainly by the residence time of pore water in soil, soil microbial activity, and content of organic matter and clay silicates.

8   Diagnosis

Dawra and Sharma (2001) described the conventional diagnostic methods that have been used. The newer diagnostic tests include gene mutation detection, immunohistochemical detection of tumor biomarkers, chromosomal aberrations, oxidative stress for EBH, and, PCR, real-time PCR and DNase-SISPA (sequence-independent single primer amplification) for BPVs.

8.1   Gene Mutations

One or more molecular biomarkers are needed to detect the carcinogenicity of bracken fern in animals. The most frequently detected alterations to oncogenes, in both animal and tumor model systems of human cancers, are mutations in the *ras* family of oncogenes (*H-ras*, *K-ras*, and *N-ras*) (Brown et al. 1990; Hoffmann et al. 1993; Krontridis et al. 1993). The *ras* gene family codes for proteins of 21 kDa (p21), which are found in the cytoplasm and are associated with the inner surface of the plasma membrane. The normal function of the p21 proteins is to interact with tyrosine kinase receptors to activate a signal transduction pathway. Therefore, all of the *ras* gene products have GTPase activity, and regulate cell growth and differentiation. The mutated *ras* p21 has a structure that hinders its ability to bind to the GTPase activating protein (GAP), thus keeping p21 in the GTP-bound activated state. Most of these mutations are point mutations in codon 12, 13, or 61, which convert the *ras* gene to a transforming oncogene (Bos 1989). Mutation screening in the *ras* genes may be a useful marker for the early detection of bladder cancer. However, the incidence of *ras* gene activation in cancers varies greatly. Tissue and organ specificities of *ras* gene activation in humans have been reported, both with respect to mutated codon and type of *ras* gene (Balmain and Brown 1988). In several tumor types, a mutant gene is only occasionally found, and in a number of tumor types, no mutated gene has thus far been identified (Przybojewska et al. 2000). Several studies have been carried out on the genetic mutations associated with human bladder cancer.

Mutations of the *p53* gene are the most frequent somatic genetic abnormalities detected in human malignant disease, and tend to be more common in urinary bladder cancer (Hainaut and Hollstein 2000; Dalbagni et al. 2001). The *p53* gene encodes a 53 kDa transcription factor with a critical role in DNA repair and apoptosis. Most of the *p53* mutations are clustered in the evolutionarily highly conserved and functionally important exons 5–8 of the gene (Olivier et al. 2002). The mutated *p53* protein has a much longer half-life than the wild-type *p53*, thus allowing its detection by immunohistochemistry. It has been shown in humans that *p53* mutations are found mainly in high-grade and invasive bladder tumors, constituting about 35–72% of muscle-invasive bladder tumors (Prescott et al. 2001; Hartmann et al. 2002; Lu et al. 2002). Moreover, most *p53* alterations are missense point mutations, which are clustered (80%) in the important exons 5–8 of the gene (Olivier et al. 2002). Recent studies indicated that the detection of a *p53* mutation, loss of heterozygosity (LOH) analysis, and overexpression of the *p53* protein provide valuable information regarding the prognosis and responsiveness to therapy of human urinary bladder carcinoma (Shigyo et al. 2001; Smith et al. 2003). Thus, detection of any mutations in the *ras* and *p53* genes in bracken fern-induced carcinogenesis will help to better understand the mechanism involved in carcinogenesis.

However, limited studies have been done on bracken fern-induced tumor in animals. Prakash et al. (1996) reported evidence of PT-DNA adducts and mutations at codon 61 of the *H-ras* gene in ileal epithelial cells of bracken fern-fed calves up to 28 days after bracken feeding had started. A transition in codon 12 and a transversion in codon 61 of this gene (Krengel et al. 1990) induced loss of GTPase activity of the proto-oncogene product, p21 protein, which led to an uncontrolled cell cycle and to tumorigenesis. There was early activation of the *H-ras* oncogene in rats from PT carcinogenesis (Shahin et al. 1998b). Subsequently, Sardon et al. (2005) conducted *H-ras* molecular analysis to detect mutations in the urinary bladders of naturally EBH-affected cattle in Spain. A silent mutation in exon 2 (D38D) of *H-ras* was detected in one animal afflicted with a mixed bladder tumor. A further polymorphism was also present in intron I (nt 1127), although there was no congruence between urinary bladder phenotypes. No mutations were found in codons 12, 59, or 61, but this does not exclude the presence of polymorphisms in other regions of the gene (promoter or regulation sequences) or in other genes (belonging or not to the ras family) that significantly affect the H-ras protein. However, Freitas et al. (2002), in a study on bracken fern-induced malignant tumors in the rat model, could not detect any mutations in *p53*, *H-ras*, and *K-ras* genes in an experimental trial conducted over for periods up to 24 months.

8.2   *Immunohistochemical Expression of Tumor Biomarkers*

Ki-67, a nuclear antigen, is chiefly associated with cell cycle events and is considered to be a specific and important marker for cell proliferation. Ki-67 normally increases in stages and grades of various tumors as infiltration depth increases (Liu et al. 2005). Cytokeratin 7 (CK 7) and cytokeratin 20 (CK 20) expressions are generally restricted to epithelia and bladder neoplasms. Ki-67, CK 7, and CK 20 have been used to diagnose an infiltrating type of transitional cell carcinoma of urinary bladder in buffalo (Deshmukh et al. 2010). The expression of these marker proteins in tissue samples increases sensitivity and discriminatory power of differential diagnosis and helps to correlate with pathologic features.

Sardon et al. (2005) examined the urinary bladders of natural EBH-affected cattle in Spain by employing *H-ras* immunohistochemistry. In a macroscopical and histological study of urinary bladder lesions found at a slaughterhouse, an incidence of chronic cystitis (34.1%) and tumors (2.4%) was revealed. In this study, the immunohistochemical expression was significantly increased of *H-ras* ($P<0.05$) in chronic cystitis (*H-ras*=53.24%) and bladder tumors (*H-ras*=63.60%), compared with normal urinary bladders (*H-ras*=4.32%). A silent mutation in exon 2 was detected in one animal with a mixed bladder tumor. No mutations were found in codons 12, 59, or 61. The immunohistochemical expression of *H-ras* indicated aberrant accumulation of the protein that was caused by mechanisms other than mutations in codons 12, 59, and 61 of the *H-ras* gene. It seemed that elevated *H-ras* immunoexpression in the inflammatory lesions and tumors indicated protein accumulation related to preneoplastic or neoplastic metabolic changes. The absence of mutations in the *H-ras* codons studied does not exclude the presence of polymorphisms in other regions of the gene (promoter or regulation sequences), or in other genes (belonging or not to the *ras* family) that significantly affect the H-ras protein. Therefore, it is necessary to analyze those sequences precisely. *H-ras* immunohistochemistry may, therefore, be valuable in identifying both early and later stages of the disease, and if employed may help avoid human exposure to bracken fern carcinogens.

Expression of factor VIII-related antigen (FVIIIra), CD31, muscle-specific actin, uroplakin III (UPIII), and the cell cycle-related proteins cyclin D1 and *p53* has been evaluated in hemangiomas, hemangioendotheliomas, and hemangiosarcomas in 26 natural EBH-affected cows in Portugal. Fifty-six endothelial-derived urinary bladder tumor samples were collected from 26 animals afflicted with EBH. Although CD31 expression was seen in all endothelial tumors tested, FVIIIra was not expressed in poorly differentiated endothelial tumor cells taken from solid areas or in seven muscle-invasive hemangiosarcomas. Cyclin D1 overexpression was seen in 53% of hemangiomas, 82% of hemangioendotheliomas, and 95% of hemangiosarcomas. *P53* immunoreactivity was only seen in muscle-invasive hemangiosarcomas. The UPIII staining pattern, normally very intense on the apical aspect and cytoplasm of superficial urothelial cells, was altered in the urothelium in an estimated 25% of hemangiomas and in most hemangioendotheliomas and hemangiosarcomas. It was concluded that CD31 is a better marker than FVIIIra for characterizing bovine endothelial tumors. The cell cycle regulatory pathways involving cyclin D1 and *p53* seem to be impaired in endothelial urinary bladder tumors, when *p53* immunoreactivity positively correlates with enhanced invasion (Carvalho et al. 2009).

8.3   Chromosomal Aberrations

Chromosomal aberrations, especially chromatid breaks, are also known to occur in human urinary bladder cancer (Bryant et al. 2004). There is little information on chromosomal instability in cattle that have access to bracken fern and suffer from EBH. Lioi et al. (2004) observed an increase in the number of structural chromosomal aberrations in 56 cattle with EBH that were raised on pastures in which they had bracken fern access. In this study, the highest clastogenic effect was observed in cattle that had urinary bladder cancer (27 animals). Subsequently, Peretti et al. (2007) investigated chromosomal aberrations (aneuploidy, gaps, chromatid breaks, chromosome breaks, and fragments) and sister chromatid exchanges (SCEs) in 45 cattle, aged 7–12 years, that were pastured in the south of Italy. Of these animals, 30 had access to bracken fern and showed signs of EBH, whereas 15 (control) did not. The percentage of abnormal cells (aneuploidy, chromatid breaks, chromosome breaks, and fragments) was higher in animals affected by EBH (34.7%) than in the control group (24.3%). The same results were achieved when gaps were included. Cytogenetic analysis revealed pronounced chromosome fragility in cells of animals affected by EBH, since a significantly higher ($P<0.001$) frequency of abnormal cells without gaps was detected in animals affected by EBH than in controls. After gaps, aneuploidy was the most common abnormality found in the animals, followed by chromatid breaks, chromosome breaks, and fragments (Peretti et al. 2007). In contrast to the earlier studies, cytogenetic analysis of natural EBH-affected animals in Uttar Pradesh, India, did not reveal any significant changes in chromosome pattern, as compared to controls (Singh 2007).

8.4 Oxidative Stress

Limited studies on oxidative stress from consuming bracken fern have been carried out in animals. The status of lipid peroxidation, glutathione, glutathione peroxidase, glutathione reductase, glutathione-S-transferase, superoxide dismutase, catalase, ascorbic acid, and α-tocopherol were studied in the urinary bladder of guinea pigs exposed to carcinogenic fern *O. contiguum* (Sood et al. 2003). There was a significant increase in the preformed lipid peroxides in the urinary bladders from fern-exposed animals. The concentrations of glutathione and α-tocopherol and the activities of glutathione reductase and catalase were elevated in the urinary bladders of the animals exposed to the fern. No effect was observed on the concentration of ascorbic acid and the activities of glutathione peroxidase, glutathione-S-transferase, and superoxide dismutase. It was concluded that the fern toxins increased oxidative stress in the urinary bladder, and antioxidant status was altered. However, the altered antioxidant status did not provide protection from the toxin-induced injury. Leishangthem (2006) performed a study on *P. aquilinum* (L) Kuhn and on the *O. contiguum* ferns-bovine papillomavirus and their interaction in Golden hamsters. Results were, after 4 months of exposure, increased glutathione-S-transferase activity in brain, liver, kidneys, ileum, and skin.

8.5 Ultrasonography

Ultrasonography (USG) has been used as a complementary diagnostic aid to evaluate bladder wall thickness and the presence of space-occupying lesions, in cases of natural and experimental EBH-associated urinary bladder tumors. In addition, USG can be utilized to detect preclinical cases of EBH and to diagnose early experimental fern-induced urinary bladder tumors, where hematuria has not developed (Hoque et al. 2002).

8.6   Polymerase Chain Reaction for Bovine Papillomavirus

It is not easy to demonstrate BPV-2 in the etiology of bovine urinary bladder carcinoma by conventional virological methods, such as virus isolation in cell culture. Therefore, the integrity of epidemiological studies relies on methods that are sensitive and specific for BPV-2 detection and typing. Moreover, PCR and multiplex PCR, seminested PCR have also been used to increase the results specificity. Nested PCR refers to a pair of PCRs run in series, each with a pair of primers flanking the same sequence. The first PCR amplifies a sequence as seen in any PCR experiment. The second pair of primers (nested primers) for the second PCR bind within the first PCR product and produce a second PCR product that is shorter than the first one. Because it uses four specific primers, rather than two, this technique has greater specificity than does regular PCR. It also yields detectable product where simple PCR fails to do so. Seminested PCR is similar to a nested PCR except that in the second PCR one of the primers is a primer that was used in the first PCR. In addition, a multiplex PCR has been evaluated to detect for the BPV-2 L1 gene and bovine mitochondrial genome ND5 gene (internal control), followed by a second round of BPV-2 amplification via a seminested PCR (SN-PCR) (Wosiacki et al. 2005).

Wosiacki et al. (2005) used six skin papilloma samples for PCR technique development. Twenty-two urinary bladder samples from symptomatic ($n=12$) and asymptomatic ($n=10$, control group) cows and 25 blood samples from cows grazed on EBH-endemic ($n=14$) and EBH-free ($n=11$, control group) pastures in geographical regions of Parana State, Brazil, were analyzed. The SN-PCR detected BPV-2 in seven urinary bladder and ten whole blood samples collected from cows with enzootic hematuria, and in one urinary bladder and one whole blood sample of asymptomatic cows. The specificity of the amplicon was performed by restriction fragment length polymorphism and sequence analysis. The SN-PCR technique developed in this study made it possible to diagnose and compare epidemiological studies in which BPV-2 infection rates were evaluated in cattle. This allowed the authors (Wosiacki et al. 2005) to establish the association of this infection with bracken fern chronic intoxication in the etiology of EBH and opened the possibility of antemortem studies by lymphocyte analysis. The development of a new diagnostic method (e.g., SN-PCR), as described in this study, would allow the evaluation of BPV-2 infection rates in cattle and the association of this infection with bracken fern chronic intoxication in EBH etiology. In addition, this technique is highly suited for detecting BPV-2 in bovine lymphocytes, thereby allowing antemortem studies to be carried out (Wosiacki et al. 2005). The SN-PCR assay has also been used to assess the presence of the BPV-2 in the urinary bladder samples that displayed hyperplastic and hemorrhagic macroscopic lesions; these samples were collected from adult cattle in geographical regions where the EBH was endemic (Wosiacki et al. 2006).

Roperto et al. (2008) investigated the simultaneous presence of BPV-2 in whole blood and urinary bladder tumors of adult cattle in an attempt to better understand the biological role of circulating BPV-2. Peripheral blood samples were collected from 78 cattle clinically suffering from a severe chronic EBH. Circulating BPV-2 DNA was detected in 61 of them and in two blood samples from healthy cows. Fifty of the affected animals were slaughtered at public slaughterhouses, and neoplastic proliferations in the urinary bladder were detected in all of them. BPV-2 DNA was amplified and sequenced in 78% of urinary bladder tumor samples, and in 38.9% of normal samples from the control group. Circulating episomal BPV-2 DNA was detected in 78.2% of the blood samples. The simultaneous presence of BPV-2 DNA in neoplastic bladder and blood samples was detected in 37 animals. Specific viral E5 mRNA and E5 oncoprotein were also detected in blood by RT-PCR and western blot/immunocytochemistry, respectively. It is likely that BPV-2 can persist and be actively maintained in the bloodstream, particularly in the lymphocytes, as a reservoir of viral infection that, in the presence of cocarcinogens, may cause the development of urinary bladder tumors.

8.7 Real-Time PCR for BPV

A rapid, sensitive, and reliable real-time SYBR Green PCR test to detect BPV-1, BPV-2 and to quantify BPV-1 has been developed by Pangty et al. (2010a). Results of amplification and dissociation plot of real-time PCR revealed that six samples were BPV-1 positive, eight were BPV-2 positive, and six were positive for both BPV-1 and -2. Cutaneous wart (CWs) samples from different dairy farms testing positive for BPV-1 by PCR assay, and were also positive using the quantitative real-time SYBR Green PCR assay. For the first time, mixed infection of BPV-1 and -2 was detected in India, and BPV-1 load was quantified by real-time SYBR Green PCR assay.

Ferritin heavy chain (FHC) is upregulated in papillomavirus-associated urothelial tumors of the urinary bladder in cattle (Roperto et al. 2010). The upregulation of FHC was reported in six papillary and in four invasive urothelial tumors of the urinary bladder of cattle grazing on Italian mountain pastures that were rich in bracken fern. All tumors contained a sequence of BPV-2 as determined by PCR analyses, and were validated by direct sequencing of the amplified products. The oncoprotein E5 was also detected in these tumors by immunoprecipitation, immunofluorescence, and laser scanning confocal microscopy. Expression of FHC was evaluated by western blot analysis, RT-PCR, real-time RT-PCR, and immunohistochemistry. The oligonucleotide sequence of the bovine ferritin amplicons was identical to that of human ferritin.

Recently, Pathania et al. (2012) evaluated the applicability of PCR and real-time PCR to detect BPV-2 in urine samples and urinary bladder lesions in bovines and quantified BPV-2 in urinary bladder lesions in EBH-affected animals. These authors tested 24 urine samples collected from cows with a clinical history of cutaneous warts (3), teat warts (3), hematuria (9), and cows that were apparently normal (9). Of these, 12 (50%) were positive for BPV-2 by PCR. These included three animals that exhibited CWs, four animals showing hematuria, and five indicating an origin from an enzootic locality. In addition, five animals with hematuria and five with a history of originating from enzootic locality were found to be negative. Among the 35 urinary bladder samples analyzed with PCR, 24 (68.6%) were positive. Of 28 urinary bladder samples, 24 (85.7%) were found to be positive by using real-time PCR, as follows: EBH-affected animals (9), acute cystitis (10), chronic cystitis (4), and bladders without microscopic lesions (1). Use of the quantitative real-time PCR (SYBR Green assay) showed that the BPV-2 load was low and was comparable, irrespective of the presence of inflammatory or neoplastic bladder lesions.

These results indicated that urine samples from endemic areas of EBH were positive for BPV-2 DNA, regardless of specific clinical history. Results revealed that 15 cases of nonneoplastic lesions and 9 cases of EBH-affected animals showed the presence of BPV-2 DNA, which confirmed widespread prevalence of BPV-2 in the urinary bladder, irrespective of lesion type. It was concluded that BPV-2 DNA is widely present in both urine and different types of urinary bladder lesions in cows from an EBH endemic region.

8.8   DNase-SISPA for BPV

Papillomaviruses are usually detected and identified by PCR, and display consensus primers designed from human papillomavirus sequences. Despite repeated attempts, these and other primers could not detect papillomavirus in bovine teat wart samples. DNase-SISPA (sequence-independent single primer amplification), a metagenomic method for identifying viruses, was used to identify bovine papillomavirus type 10 in bovine teat warts. The sequence comparison between consensus primers and bovine papillomavirus type 10 sequences revealed many differences between consensus primers and BPV-10 sequences. Rai et al. (2011) suggested that DNase-SISPA may be used as an alternate method for diagnosing papillomavirus, where PCR fails to identify papillomaviruses.

9   Prevention and Control

Campos-da-Paz et al. (2008) studied the in vitro effect of vitamin C on the reversibility of DNA damage caused by bracken fern on human submandibular gland (HSG) cells and on oral epithelium cells (OSCC-3) that had previously been exposed to bracken fern extract. They found that vitamin C (10 μg/mL) alone did not reduce DNA damage caused by bracken fern in HSG and OSSC-3 cells. However, at a higher concentration (100 μg/mL), vitamin C did induce DNA damage in both cell lines. Moreover, vitamin C (10 and 100 μg/mL), together with bracken fern extract, showed a synergistic effect on the frequency of DNA damage in HSG cells. DNA damage (i.e., nuclei with increased levels of DNA migration) was determined by comet assay, cell morphology was evaluated by light microscopy, and cellular degeneration was assessed by the acridine orange/ethidium bromide fluorescent-dyeing test.

Latorre et al. (2011) showed that selenium supplementation could prevent and reverse the immunotoxic effect of PT. It has already been shown that bracken fern (*P. aquilinum*) has immunomodulatory effects on mouse natural killer (NK) cells, and acts by reducing cytotoxicity. Alternatively, it has been demonstrated that selenium enhances NK cell activity. In this study, male C57BL/6 mice were administered the *P. aquilinum* extract by daily gavage for 30 days, and histological analyses revealed a significant reduction in the splenic white pulp area that was fully reversed by selenium treatment. In addition, mice administered PT by daily gavage for 14 days demonstrated the same reduction of NK cell activity as the *P. aquilinum* extract, and this reduction was prevented by selenium coadministration. Lastly, nonadherent splenic cells treated in vitro with an RPMI (Roswell Park Memorial Institute cell culture medium) extract of *P. aquilinum* also showed diminished NK cell activity that was not only prevented by selenium co-treatment, but was also fully reversed by selenium posttreatment.

In India, a preliminary prophylactic and therapeutic vaccine trial was successfully conducted in bull calves. The treatment administered was a binary ethylenimine (BEI)-inactivated saponized crude cutaneous warts bovine papillomavirus 1 and 2 (CW BPV-1 and -2) vaccine. In a prophylactic group of vaccinated animals, no temperature effect, untoward reaction, or nodule formation was seen at the vaccine inoculation site. After challenge with BPV inoculum, cutaneous warts (CWs) failed to develop in this group, indicating that the vaccine was effective. In the therapeutic group, CWs were experimentally induced in two vaccinated animals. In one animal, luxuriant cauliflower-like and in other linear slow growth was observed. After vaccination, CWs in both animals became dried and growth was arrested. It appeared that tumors were regressing, which was confirmed by further clinical observations as very little tumor-scaring was left at the termination of the experiment. Histopathologically, the lesions in both animals were diagnosed as regressing occult/early fibroblastic type papillomas. Mononuclear cellular infiltration/aggregates around hair follicles and in tumor stroma were noted. The mononuclear cellular inflammatory reaction was more pronounced in the first tumors (those that were cauliflower-like and fast growing), than those appearing in the second case (linear growth/fast regressing). The clinical regression results were in accord with histopathological findings, relevant to the degree of infiltration of lymphocytes; this trial showed that in both types, use of BEI-inactivated saponized crude CW BPV-1 and BPV-2 vaccine gave encouraging results, in that CWs either quickly regressed or were prevented (Pangty et al. 2010b).

Recently, Pathania et al. (2011) conducted a preliminary therapeutic vaccine trial in hill cows to evaluate the therapeutic potential of binary ethylenimine

(BEI)-inactivated and saponized bovine papillomavirus-2 (BPV-2) for EBH. Although the vaccine failed to show favorable clinical vaccine results for EBH-affected cows at 120 days postvaccination, the immunopathological responses were encouraging. A significant difference was observed in humoral and cell-mediated immune responses following vaccination. The vaccinated animals grossly failed to show regression of bladder tumors, but microscopically showed engorgement and marked perivascular infiltration of mononuclear cells, which denote the initial stages of tumor regression. Overall, results indicated that the therapeutic vaccine developed has potential for treating EBH in cows, although further modifications in vaccine dose and field trials of the vaccine are needed.

10 Future Areas for Research

The areas ripe for future research include fern-induced carcinogenesis, nature of carcinogens involved, disease etiology, and interaction of fern carcinogens with papillomavirus. We suggest the following areas be emphasized:

1. Additional research is needed to identify PT and other putative carcinogens in other fern species and to assess the carcinogenic potential in a more complete cross section of fern flora.
2. Work is needed on the many other fern species that grow in pastures in enzootic areas, in addition to bracken (Dawra et al. 1989; Somvanshi et al. 2006; Nagarajan et al. 2011). Most research on bracken fern to date has addressed urinary bladder carcinogenesis or closely related topics. Only limited research has been performed to either analyze for PT in other ferns (Nagarajan et al. 2011) or to assess potential to induce carcinogenicity as bracken does (Dawra et al. 2001).
3. Long-term experimental studies are needed to address the interaction between BPV2 and fern carcinogens. The existing literature has produced no clear conclusions on this topic. Moreover, an assessment is needed on the status of papillomavirus in enzootic areas.
4. Analytical and epidemiological data on human health is needed on a priority basis in enzootic areas, because the effect of fern carcinogens on human health has received limited attention. The possibility that PT or other putative fern carcinogens enters the human food chain exists in enzootic areas around the world.
5. Although studies on the entry of PT into soil and water channels have added another dimension to effect of bracken and other ferns to the overall environmental effects, future work is needed to identify and evaluate contaminated soil and water sources. Cocktail effects and covariation should be included in any future environmental studies with PT.
6. Although limited information is available on the mechanism of PT-induced carcinogenesis, work is needed to define the detailed molecular mechanism by which PT and other potential carcinogens induce carcinogenesis. Knowledge of

the exact PT-induced molecular gene mutations will help develop early diagnostic methods for EBH.

7. Finally, a specific treatment for the condition is still lacking; future research should be focused on early diagnosis and development of methods for interrupting the carcinogenesis process.

11 Summary

In this review, we address the major aspects of enzootic bovine hematuria and have placed special emphasis on describing the etiology, human health implications, and advanced molecular diagnosis of the disease.

Enzootic bovine hematuria (EBH) is a bovine disease characterized by the intermittent presence of blood in the urine and is caused by malignant lesions in the urinary bladder. This incurable disease is a serious malady in several countries across many continents. Accurate early-stage diagnosis of the disease is possible by applying advanced molecular techniques, e.g., detection of genetic mutations in the urine of cows from endemic areas. Use of such diagnostic approaches may help create an effective therapy against the disease.

There is a consensus that EBH is caused primarily by animals consuming bracken fern (*P. aquilinum*) as they graze. The putative carcinogen in bracken is ptaquiloside (PT), a glycoside. However, other bracken constituents like quercetin, isoquercetin, ptesculentoside, caudatoside, astragalin, and tannins may also be carcinogenic. Studies are needed to identify the role of other metabolites in inducing urinary bladder carcinogenesis.

The bovine papillomavirus is also thought to be an associated etiology in causing EBH in cattle. There is growing alarm that these fern toxins and their metabolites reach and contaminate the soil and water environment and that the carcinogen (PT) is transmitted via cow's milk to the human food chain, where it may now pose a threat to human health. An increased incidence of gastric and esophageal cancer has been recorded in humans consuming bracken ferns, and among those living for long periods in areas infested with bracken ferns.

Although preliminary therapeutic vaccine trials with inactivated BPV-2 against EBH have been performed, further work is needed to standardize and validate vaccine doses for animals.

References

Agnew MP, Lauren DR (1991) Determination of ptaquiloside in Bracken fern (*Pteridium esculentum*). J Chromatogr 538:462–468

Alonso-Amelot ME (1997) The link between bracken fern and stomach cancer: milk. Nutrition 13:694–696

Alonso-Amelot ME, Avendano M (2002) Human carcinogenesis and bracken fern: a review of the evidence. Curr Med Chem 9:675–686

Alonso-Amelot ME, Rodulfo-Baechler S, Jaimes-Espinosa R (1995) Comparative dynamics of ptaquiloside and pterosin B in the two varieties (*caudatum* and *arachnoideum*) of neotropical bracken fern (*P. aquilinum* L. Kuhn). Biochem Syst Ecol 23:709–716

Alonso-Amelot ME, Castillo U, Smith BL, Lauren DR (1998) Excretion, through milk, of pta-quiloside in bracken-fed cows—a quantitative assessment. Lait 78:413–423

Aparecida da Silva M, Scardua CM, Dorea MD, Carvalho NL, Martins IVF, Donatele DM (2009) Prevalence of bovine enzootic haematuria in dairy cattle in the Caparao microregion, southern Espirito Santo, between 2007 and 2008. Ciencia Rural, Santa Maria 39:1847–1850

Ayala-Luis KB, Hansen PB, Rasmussen LH, Hansen HCB (2006) Kinetics of ptaquiloside hydro-lysis in aqueous solution. Environ Toxicol Chem 25:2623–2629

Balcos LGF, Borzacchiello G, Russo V, Popescu O, Roperto S, Roperto F (2008) Association of bovine papillomavirus type-2 and urinary bladder tumours in cattle from Romania. Res Vet Sci 85:145–148

Balmain A, Brown K (1988) Oncogene activation in chemical carcinogenesis. Adv Cancer Res 51:147–182

Barbacid M (1986) Oncogenes and human cancer: cause or consequence? Carcinogenesis 7:1037

Basaran AA, Yu T-W, Plewa MJ, Anderson D (1996) An investigation of some Turkish herbal medicines in *Salmonella typhimurium* and in the comet assay in human lymphocytes. Teratog Carcinog Mutagen 16:125–138

Borzacchiello G, Ambrosio V, Galati P, Perillo A, Roperto F (2003a) Cyclooxygenase-1 and -2 expression in urothelial carcinomas of the urinary bladder in cows. Vet Pathol 40:455–459

Borzacchiello G, Iovane G, Marcante ML, Poggiali F, Roperto F, Roperto S, Venuti A (2003b) Presence of BPV-2 DNA and expression of the viral oncoprotein E5 in naturally occurring bladder tumours in cows. J Gen Virol 84:2921–2926

Borzacchiello G, Russo V, Spoleto C, Roperto S, Balcos L, Rizzo C, Venuti A, Roperto F (2007) Bovine papillomavirus type-2 DNA and expression of E5 and E7 oncoproteins in vascular tumours of the urinary bladder in cattle. Cancer Lett 250:82–89

Bos JL (1989) Ras oncogenes in human cancer: a review. Cancer Res 49:4682–4689

Brandt S, Haralambus R, Schoster A, Kirnbauer R, Stanek C (2008) Peripheral blood mononuclear cells represent a reservoir of bovine papillomavirus DNA in sarcoid-affected equines. J Gen Virol 89:1390–1395

Bringuier PP, Piaton E, Berger N, Debruyne F, Perrin P, Schalken J, Devonec M (1995) Bracken fern-induced bladder tumors in guinea pigs—a model for human neoplasia. Am J Pathol 147:858–868

Brown K, Buchmann A, Balmain A (1990) Carcinogen induced mutations in the mouse c-*Ha-ras* gene provide evidence of multiple pathways for tumor progression. Proc Natl Acad Sci U S A 87:538–542

Brun R, Urraro C, Medaglia C, Russo V, Borzacchiello G et al (2008) Lymphoepithelioma-like carcinoma of the urinary bladder in a cow associated with bovine papillomavirus type-2. J Comp Pathol 139:121–125

Bryant PE, Gray LJ, Peresse N (2004) Progress towards understanding the nature of chromatid breakage. Cytogenet Genome Res 104:65–71

Campo MS (1997) Bovine papillomavirus and cancer. Vet J 154:175–188

Campo MS (2006) Bovine papillomavirus: old system, new lessons? In: Campo MS (ed) Papillomavirus research from natural history to vaccines and beyond. Caister Academic Press, Norfolk, pp 373–383

Campo MS, McCafferty RE, Doherty I, Kennedy IM, Jarrett WFH (1990) The Harvey ras 1 gene is activated in papillomavirus-associated carcinomas of the upper alimentary canal in cattle. Oncogene 5:303–308

Campo MS, Jarrett WFH, Barron R, O'Neill BW, Smith KT (1992) Association of bovine papil-lomavirus type 2 and bracken fern with bladder cancer in cattle. Cancer Res 52:6898–6904

Campos-da-Paz M, Pereira LO, Bicalho LS, Dorea JG, Pocas-Fonseca MJ, Almeida Santos MFM (2008) Interaction of bracken-fern extract with vitamin C in human submandibular gland and oral epithelium cell lines. Mutat Res 652:158–163

Carvalho T, Pinto C, Peleteiro MC (2006) Urinary bladder lesions in bovine enzootic haematuria. J Comp Pathol 134:336–346

Carvalho T, Naydan D, Nunes T, Pinto C, Peleteiro MC (2009) Immunohistochemical evaluation of vascular urinary bladder tumors from cows with enzootic hematuria. Vet Pathol 46:211–221

Castillo UF, Ojika M, Sakagami Y, Wilkins AL, Lauren DR, Alonso-Amelot ME, Smith BL (2000) Isolation and structural determination of three new toxic illudane-type sesquiterpene glucosides and a new protoilludane sesquiterpene glucoside from Pteridium aquilinum var. caudatum. In: Taylor JA, Smith RT (eds) Bracken fern: toxicity, biology and control. Proceedings of the international bracken group conference Manchester. Aberystwyth: International Bracken Group, UK, pp 55–59

Chambers G, Ellsmore VA, O'Brien PM, Reid SW, Love S, Campo MS, Nasir L (2003) Association of bovine papillomavirus with the equine sarcoid. J Gen Virol 84:1055–1062

Costa Rui MG, Oliveira PA, Vilanova M, Bastos MMSM, Lopes CC, Lopes C (2011) Ptaquiloside-induced, B-cell lymphoproliferative and early-stage urothelial lesions in mice. Toxicon 58:543–549

Dalbagni G, Ren ZP, Herr H, Cordon-Cardo C, Reuter V (2001) Genetic alterations in tp53 in recurrent urothelial cancer: a longitudinal study. Clin Cancer Res 7:2797–2801

Dawra RK, Sharma OP (2001) Enzootic bovine haematuria—past, present and future. Vet Bull 71:1–27

Dawra RK, Sharma OP, Krishna L (1989) Are environmental conditions responsible for animal-plant carcinogen interaction? A study relating to enzootic bovine haematuria. Environmentalist 9:277–283

Dawra RK, Sharma OP, Krishna L, Vaid JL (1991) The enzymatic profile of urine and plasma in bovine urinary bladder cancer (enzootic bovine haematuria). Vet Res Commun 15:421–426

Dawra RK, Sharma OP, Somvanshi R (2001) A preliminary study on the carcinogenicity of the common fern Onychium contiguum. Vet Res Commun 25:413–420

Dawra RK, Kurade NP, Sharma OP (2002) Carcinogenicity of the fern Pteridium aquilinum collected from enzootic bovine haematuria-free hilly area in India. Curr Sci 83:1005–1009

de Villiers EM, Fauquet C, Broker TR, Bernard HU, zur Hausen H (2004) Classification of papillomaviruses. Virology 324:17–27

Deshmukh S, Banga HS, Kwatra KS, Singh ND, Gadhave PD, Brar RS (2010) Immunohistochemical study on spontaneous transitional cell carcinoma of urinary bladder in an Indian water buffalo (Bubalus bubalis). Indian J Vet Pathol 34:113–116

Engel P, Brandt KK, Rasmussen LH, Ovesen RG, Sorensen J (2007) Microbial degradation and impact of Bracken toxin ptaquiloside on microbial communities in soil. Chemosphere 67:202–209

Epstein JL, Amin MB, Reuter VR (2004) Bladder biopsy interpretation. Lippincott Williams & Wilkins, Philadelphia, PA, pp 11–34

Farajzadeh H, Tehrani AA, Amirarfaei Y, Pourata A, Mansoub NH, Khalili Y, Mohammadi M (2011) Pathological studies on the incidence of bracken-fern induced encephalomalacia in rat. Ann Biol Res 2:386–392

Fenwick GR (1988) Bracken (Pteridium aquilinum)—toxic effects and toxic constituents. J Sci Food Agric 46:147

Fletcher MT, Brock IJ, Reichmann KG, McKenzie RA, Blaney BJ (2011) Norsesquiterpene glycosides in bracken ferns (Pteridium esculentum and Pteridium aquilinum subsp. wightianum) from Eastern Australia: reassessed poisoning risk to animals. J Agric Food Chem 59:5133–5138

Freitas RN, O'Connor PJ, Prakash AS, Shahin M, Povey AC (2001) Bracken (Pteridium aquilinum)-induced DNA adducts in mouse tissues are different from the adduct induced by the activated form of the bracken carcinogen ptaquiloside. Biochem Biophys Res Commun 281:589–594

Freitas RN, Brasileiro-Filho G, Silva ME, Pena SDJ (2002) Bracken fern-induced malignant tumors in rats: absence of mutations in p53, H-ras and K-ras and no microsatellite instability. Mutat Res 499:189–196

Gangwar NK (2004) Studies on pathological effects of *linguda* (*Diplazium esculentum, Retz.*) in laboratory rats and guinea pigs. M.V.Sc. Thesis, IVRI Deemed University, Izatnagar-243122, Bareilly, India

Gerenutti M, Spinosa HS, Bernardi MM (1992) Effects of bracken fern (*Pteridium aquilinum* L Kuhn) feeding during the development of female rats and their offspring. Vet Human Toxicol 34:307–310

Giurgiu G, Mircean M, Scurtu I, Popovici C (2008) Epidemiological, clinical and paraclinical investigations in cattle enzootic haematuria in an area from N-W of Transilvania. Bull Univ Agric Sci Vet Med 65:12–15

Gounalan S, Somvanshi R, Kataraia M, Bisht GS, Smith BL, Lauren DR (1999) Effect of bracken (*Pteridium aquilinum*) and dryopteris (*Dryopteris juxtaposita*) fern toxicity in laboratory rabbits. Indian J Exp Biol 37:980–985

Hainaut P, Hollstein M (2000) p53 and human cancer: the first ten thousand mutations. Adv Cancer Res 77:81–137

Hartmann A, Schlake G, Zaak D, Hungerhuber E, Hofstetter A, Hofstaedter F, Knuechel R (2002) Occurrence of chromosome 9 and p53 alterations in multifocal dysplasia and carcinoma in situ of human urinary bladder. Cancer Res 62:809–818

Hirayama T (1979) Diet and cancer. Nutr Cancer 1:67–81

Hirono C, Shibuya M, Shizmu K, Fushmi K (1972) Carcinogenic activity of processed bracken as human food. J Natl Cancer Inst 48:1245–1250

Hoffmann JS, Fry M, Williams J, Loeb LA (1993) Codon 12 and 13 of *H-ras* protooncogene interrupt the progression of DNA synthesis catalyzed by DNA polymerase alpha. Cancer Res 53:2895–2900

Hopkins NCG (1986) Aetiology of enzootic haematuria. Vet Rec 118:715–717

Hopkins NC (1987) Enzootic haematuria in Nepal. Trop Anim Health Prod 19:159–164

Hoque M, Somvanshi R, Singh GR, Mogha IV (2002) Ultrasonographic evaluation of urinary bladder in normal, fern fed and enzootic bovine haematuria-affected cattle. J Vet Med A 49:403–407

Jensen PH, Jacobsen OS, Hansen HC, Juhler RK (2008) Quantification of ptaquiloside and pterosin B in soil and groundwater using liquid chromatography-tandem mass spectrometry (LC-MS/MS). J Agric Food Chem 56:9848–9854

Jones TC, Hunt RO (1983) Veterinary pathology, 5th edn. K. M. Varghese Co., Bombay, pp 1500–1501

Jones PA, Buckley JD, Henderson BE, Ross RK, Pike MC (1991) From gene to carcinogen: a rapidly evolving field in molecular epidemiology. Cancer Res 51:3617–3620

Kerrigan J (1926) Chronic haematuria affecting cattle. Note on cases in the South Island west coast district. N Z J Agric 33:85–87

Krengel U, Schlichting I, Schere A, Schuman R, Frech M, John U, Kabish W, Pai E, Wittinghoffer A (1990) Three dimensional structures of H-ras p21 mutants: molecular basis for their inability to function as signal switch molecules. Cell 62:539–548

Krontridis TG, Devlin B, Karp DD, Robert NJ, Rish N (1993) An association between the risk of cancer and mutations in the *HRAS1* minisatellite locus. N Engl J Med 329:517–523

Kumar AK, Kataria M, Somvanshi R, Kumar S, Saini M (2001) Characterization of toxin from *Cheilanthes* fern and its effect on lymphocyte proliferation and DNA fragmentation. Indian J Exp Biol 39:1065–1067

Kushida T, Uesugi M, Sugiura Y, Kigoshi H, Tanaka H, Hirokawa J, Ojika M, Yamada K (1994) DNA damage by ptaquiloside, a potent bracken carcinogen: detection of selective strand breaks and identification of DNA cleavage products. J Am Chem Soc 116:479–486

Latorre AO, Caniceiro BD, Wysocki HL Jr, Haraguchi M, Gardner DR, Gorniak SL (2011) Selenium reverses Pteridium aquilinum-induced immunotoxic effects. Food Chem Toxicol 49:464–470

Leishangthem GD (2006) Etio-pathological studies on bovine papillomatosis with special reference to its combined effects on fern fed hamsters. M.V.Sc. Thesis, Deemed University, IVRI, Izatnagar-243 122

Leishangthem GD, Somvanshi R, Tiwari AK (2008) Detection of bovine papillomaviruses in cutaneous warts/papillomas in cattle. Indian J Vet Pathol 32:15–20

Lioi MB, Barbieri R, Borzacchiello G, Dezzi S, Roperto S, Santoro A, Russo V, Roperto F (2004) Chromosome aberration in cattle with chronic enzootic haematuria. J Comp Pathol 113:233–236

Liu R, Xu Y, Chang J (2005) The expression and significance of KA 1 and ki-67 in bladder transitional cell carcinoma. Chin Clin Oncol 2:888–893

Lopez-Beltran A, Montironi R (2004) Non-invasive urothelial neoplasms: according to the most recent WHO classification. Eur Urol 46:170–176

Lu ML, Wikman F, Orntoft TF, Charytonowicz E, Rabbani F, Zhang Z (2002) Impact of alterations affecting the p53 pathway in bladder cancer on clinical outcome, assessed by conventional and array based methods. Clin Cancer Res 8:171–179

Lucena RB, Rissi DR, Kommers GD, Pierezan F, Oliveira-Filho JC, Macedo JTSA, Flores MM, Barros CSL (2011) A retrospective study of 586 tumours in Brazilian cattle. J Comp Pathol 145:20–24

Marliere CA, Wathern P, Freitas SN, Castro MCFM, Galvao MAM (2000) Bracken fern (*Pteridium aquilinum*) consumption and oesophageal and stomach cancer in the Ouro Preto region, Minas Gerais, Brazil. In: Taylor JA, Smith RT (eds) Bracken fern, toxicity, biology and control. International Bracken Group Special Publication No 4, Manchester, UK, p 144

Marliere CA, Wathern P, Castro MCFM, O'Connor P, Galvao MA (2002) Bracken fern (*Pteridium aquilinum*) ingestion and oesophageal and stomach cancer. IARC Sci Publ 156:379–380

Marrero E, Bulnes C, Sanchez LM, Palenzuela I, Stuart R, Jacobs F, Romero G (2001) *Pteridium aquilinum* (bracken fern) toxicity in cattle in the humid Chaco of Tarija, Bolivia. Vet Human Toxicol 43:156–158

Marrero E, Bulnes C, Sanchez LM, Palenzuela J, Stuart R, Jacobs F, Romero J (2004) Chronic toxicity in cattle due to Pteridium aquilinum (bracken fern) in Tarija Department, Bolivia: an interdisciplinary investigation, Chapter 36. In: Acamovic T, Stewart CS, Pennycott TW (eds) Poisonous plants and related toxins. CABI Publishing, Wallingford, pp 248–252

Masui T, Mann AM, Macatee TL, Okamura T, Garland EM, Fujui H, Pelling JC, Cohen SM (1991) H-ras mutations in rat urinary bladder carcinomas induced by nirofurylthiazolyl formamide and sodium saccharin, sodium ascorbate, and related salts. Cancer Res 51:3471–3475

Matsuoka A, Hirosawa A, Natori S, Iwasaki S, Sofuni T, Ishidate M Jr (1989) Mutagenicity of PT, the carcinogen in bracken, and its related illudane-type sesquiterpenes. II. Chromosomal aberration tests with cultured mammalian cells. Mutat Res 215:179–185

Mori H, Sugie S, Hirono I, Yamada K, Niwa H, Ojika M (1985) Genotoxicity of ptaquiloside, a bracken carcinogen, in the hepatocytes primary culture/DNA-repair test. Mutat Res 143:75–78

Nagarajan N, Kumar P, Punetha N, Jensen DJ, Lauren DR, Somvanshi R (2011) Detection of ptaquiloside in certain ferns of Uttarakhand and Tamil Nadu, India. Proceed Ann Congress Vet Pathologists, India, 182

Niwa H, Ojika M, Wakamatsu K, Yamada K, Hirono I, Matsushita K (1983) Ptaquiloside, a novel norsesquiterpene glucoside from bracken, *Pteridium aquilinum* var. *latiusculum*. Tetrahedron Lett 24:4117–4120

Ojika M, Wakamatsu K, Niwa H, Yamasa K (1987) Ptaquiloside, a potent carcinogen isolated from bracken fern *Pteridium aquilinum* var *latiusculum*: structure elucidation based on chemical and spectral evidence, and reactions with aminoacids, nucleosides and nucleotides. Tetrahedron 43:5261–5274

Ojika M, Sugimoto K, Okazaki T, Yamada K (1989) Modification and cleavage of DNA by ptaquiloside. A new potent carcinogen isolated from bracken fern. J Chem Soc Chem Commun 22:1775

Olivier M, Eeles R, Hollstein M, Khan MA, Harris CC, Hainaut P (2002) The IARC TP53 database: new online mutation analysis and recommendations to users. Human Mutat 19:607–614

Ovesen RG, Rasmussen LH, Hansen HC (2008) Degradation kinetics of ptaquiloside in soil and soil solution. Environ Toxicol Chem 27:252–259

Pangty K, Singh S, Goswami R, Saikumar G, Somvanshi R (2010a) Detection of BPV-1 and -2 and quantification of BPV-1 by Real time PCR in cutaneous warts in cattle and buffaloes. Transbound Emerg Dis 57:185–196

Pangty K, Singh S, Pandey AB, Somvanshi R (2010b) Preliminary binary ethylenimine inactivated bovine papillomavirus (BPV) vaccine trial against cutaneous warts in bull calves: a pathological assessment. Braz J Vet Pathol 3:105–111

Pathania S, Kumar P, Leishangthem GD, Kumar D, Dhama K, Somvanshi R (2011) Preliminary assessment of binary ethylenimine inactivated and saponized cutaneous warts (BPV-2) therapeutic vaccine for enzootic bovine haematuria in hill cows. Vaccine 29(43):7296–7302. doi:10.1016/j.vaccine.2011.07.065

Pathania S, Dhama K, Saikumar G, Shahi S, Somvanshi R (2012) Detection and quantification of bovine papilloma virus type 2 (BPV-2) by real-time PCR in urine and urinary bladder lesions in enzootic bovine haematuria (EBH)-affected cows. Transbound Emerg Dis 59:79–84

Pennie WD, Campo MS (1992) Synergism between BPV-4 and the flavonoid quercetin in cell transformation in vitro. Virology 190:861–865

Peretti V, Ciotola F, Albarella S, Russo V, Di Meo GP, Lannuzzi L, Roperto F, Barbieri V (2007) Chromosome fragility in cattle with chronic enzootic haematuria. Mutagenesis 22:317–320

Perez-Alenza MD, Blanco J, Sardon D, Sanchez Moreiro MA, Rodriguez-Bertos A (2006) Clinicopathological findings in cattle exposed to chronic bracken fern toxicity. N Z Vet J 54:185–192

Pinto C, Januario T, Geraldes M, Machado J, Lauren DR, Smith BL, Robinson RC (2004) Bovine enzootic haematuria on São Miguel Island, Chapter 84. In: Acamovic T, Stewart CS, Pennycott TW (eds) Poisonous plants and related toxins. CABI Publishing, Wallingford, UK, pp 564–574

Prakash AS, Pereira TN, Smith BL, Shaw G, Seawright AA (1996) Mechanism of bracken fern carcinogenesis: evidence for H-*ras* activation via initial adenine alkylation by ptaquiloside. Nat Toxins 4:221–227

Prescott JL, Montie J, Pugh TW, McHugh T, Veltri RW (2001) Clinical sensitivity of p53 mutation detection in matched bladder tumour, bladder wash, and voided urine specimens. Cancer 91:2127–2135

Przybojewska B, Jagiello A, Jalmuzna P (2000) *H-RAS, K-RAS*, and *N-RAS* gene activation in human bladder cancers. Cancer Genet Cytogenet 121:73–77

Rai GK, Saxena M, Singh V, Somvanshi R, Sharma B (2011) Identification of bovine papilloma virus 10 in teat warts of cattle by DNase-SISPA. Vet Microbiol 147:416–419

Rajasekaran D, Somvanshi R (2001) Evaluation of pathological effects of *Cheilanthes farinosa, Onychium contiguum* ferns and their combined effect in rats. J Lab Med 2:33–40

Rajasekaran D, Somvanshi R, Dawra RK (2004) Clinical and haematological studies on *Cheilanthes farinosa, Onychium contiguum* and their combined effects in laboratory rats. Indian J Anim Sci 74:11–14

Rasmussen LH (2003) Occurrence and fate of a Bracken (*Pteridium* sp.) toxin in terrestrial environments. Ph.D. Thesis, Chemistry Department, The Royal Veterinary and Agricultural University DK-1871 Frederiksberg C, Denmark

Rasmussen LH, Hansen HCB (2003) Growth of Bracken in 953 Denmark and the content of ptaquiloside in fronds, 954 rhizomes and roots. In: Acamovic T, Stewart CS, Pennycott TW (eds) Poisonous plants and related 956 toxins. CAB International, Wallingford, UK, p 955

Rasmussen LH, Jensen LS, Hansen HCB (2003a) Distribution of the carcinogenic terpene ptaquiloside in bracken fronds, rhizomes (*Pteridium aquilinum*), and litter in Denmark. J Chem Ecol 29:771–778

Rasmussen LH, Kroghsbo S, Frisvad JC, Hansen HCB (2003b) Occurrence of the carcinogenic bracken constituent ptaquiloside in fronds, topsoils and organic soil layers in Denmark. Chemosphere 51:117–127

Rasmussen LH, Hansen HCB, Lauren DR (2005) Sorption, degradation, and mobility of ptaquiloside, a carcinogenic bracken (*Pteridium* sp.) constituent, in the soil environment. Chemosphere 58:823–835

Rasmussen LH, Lauren DR, Smith BL, Hansen HCB (2008) Variation in ptaquiloside content in bracken (*Pteridium esculentum* (Forst. f) Cockayne) in New Zealand. N Z Vet J 56:304–309

Ravisankar R (2004) Studies on long term pathological effects of bracken fern (*Pteridium aquilinum*) in laboratory rats and hill cattle. Ph.D. Thesis, Deemed University, Indian Veterinary Research Institute, Izatnagar-243 122 (UP), India

Ravisankar R, Somvanshi R, Tandon SK, Naik A (2005) Behavioural and pathological studies on laboratory rats fed with crude extract of bracken fern. Toxicol Int 12:125–128

Reddy MV, Randerath K (1987) ^{32}P-postlabeling assay for carcinogen-DNA adducts: nuclease P1-mediated enhancement of its sensitivity and applications. Environ Health Perspect 76:41

Resendes AR, Roperto S, Trapani F, Urraro C, Rodrigues A, Roperto F, Borzacchiello G (2011) Association of BPV-2 and urinary bladder tumours in cattle from the Azores archipelago. Res Vet Sci 90:526–529

Roperto S, Roberto B, Paolini F, Urraro C, Russo V, Borzacchiello G, Pagnini U, Raso C, Rizzo C, Roperto F, Venuti A (2008) Detection of bovine papillomavirus type 2 in the peripheral blood of cattle with urinary bladder tumours: possible biological role. J Gen Virol 89:3027–3033

Roperto S, Borzacchiallo G, Brun R, Leonardi L, Maiolino P, Martano M, Paciello O, Papparella S, Restucci B, Russo V, Salvatore G, Urraro C, Roperto F (2010) A review of bovine urothelial tumours and tumour-like lesions of the urinary bladder. J Comp Pathol 142:95–108

Saito K, Nagao T, Matoba M, Koyama K, Natori S, Murakami T, Saiki Y (1989) Chemical assay of ptaquiloside, the carcinogen of *Pteridium aquilinum*, and the distribution of related compounds in the pteridaceae. Phytochemistry 28:1605–1611

Saito K, Nagao T, Takatsuki S, Koyama K, Natori S (1990) The sesquiterpenoid carcinogen of bracken fern and some analogues, from the pteridaceae. Phytochemistry 29:1475–1479

Santos RC, Brasileiro-Filho G, Hojo ES (1987) Induction of tumors in rats by bracken fern (*Pteridium aquilinum*) from Ouro Preto (Minas Gerais, Brazil). Braz J Med Biol Res 20:73–77

Santos RC, Brasileiro-Filho G, Silva ME (1992) Tumorigenicity of boiling water extract of bracken fern (*P. aquilinum* (L.) Kuhn). Ciência e Tecnologia de Alimentos 12:72–76

Santos RCS, Lindsey CJ, Ferraz OP, Pinto JR, Mirandola RS, Benesi FJ, Birgel EH, Pereira CAB, Becak W (1998) Bovine *Papillomavirus* transmission and chromosomal aberrations: on experimental model. J Gen Virol 79:2127–2135

Sardon D, de la Fuente I, Calonge E, Perez-Alenza MD, Castano M, Dunner S, Pena L (2005) H-ras immunohistochemical expression and molecular analysis of urinary bladder lesions in grazing adult cattle exposed to bracken fern. J Comp Pathol 132:195–201

Schmidt B, Rasmussen LH, Svendsen GW, Ingerslev F, Hansen HCB (2005) Genotoxic activity and inhibition of soil respiration by ptaquiloside, a bracken fern carcinogen. Environ Toxicol Chem 24:2751–2756

Shahin M, Moore MR, Worrall S, Smith BL, Seawright AA, Prakash AS (1998a) H-ras activation is an early event in the ptaquiloside induced carcinogenesis: comparison of acute and chronic toxicity in rats. Biochem Biophys Res Commun 250:491–497

Shahin M, Smith BL, Worral S, Moore MR, Seawright AA, Prakash AS (1998b) Bracken fern carcinogenesis: multiple intravenous doses of activated ptaquiloside induce DNA adducts, monocytosis, increased TNFa levels and mammary gland carcinoma in rats. Biochem Biophys Res Commun 244:192–197

Shahin M, Smith BL, Prakash AS (1999) Bracken carcinogens in the human diet. Mutat Res 443:69

Shigyo M, Sugano K, Tobisu K, Tsukamoto T, Sekiya T, Kakizoe T (2001) Molecular follow up of newly diagnosed bladder cancer using urine samples. J Urol 166:1280–1285

Siman SE, Povey AC, Ward TH, Margison GP, Sheffield E (2000) Fern spore extracts can damage DNA. Br J Cancer 83:69–73

Singh V (2007) Etio-pathological studies on bovine papillomatosis and enzootic bovine haematuria with special reference to combined effects of BPV and fern in hamsters. M.V.Sc. Thesis submitted to Deemed University, IVRI, Izatnagar-243 122, UP, India

Sivasankar S, Somvanshi R (2001) Pathological evaluation of *Polystichum squarrosum* (D. Don) fern in laboratory rats Indian. J Exp Biol 39:772–776

Smith BL, Beatson NS (1970) Bovine enzootic haematuria in New Zealand. N Z Vet J 18:115–120

Smith BL, Embling PP, Agnew MP, Lauren DR, Holland PT (1988) Carcinogenicity of bracken fern (*Pteridium esculentum*) in New Zealand. N Z Vet J 36:56–58

Smith BL, Embling PP, Lauren DR, Agnew MP, Ross AD, Greentree PL (1989) Carcinogen in rock fern (Chelianthus sieberi) from New Zealand and Australia. Aust Vet J 66:154–155

Smith BL, Embling PP, Lauren DR, Agnew MP (1992) Carcinogenicity of Pteridium esculentum and Cheilanthes sieberi in Australia and New Zealand. In: James LF, Keeler RF, Bailey EM, Cheeke PR, Hegarty MP (eds) Poisonous plants. Proceedings of the third international symposium. Iowa State University Press, Ames, IA, p 448

Smith BL, Seawright AA, Ng JC, Hertle AT, Tomson JA, Bostock PD (1994a) Concentration of ptaquiloside, a major carcinogen in bracken fern (*Pteridium* spp.), from eastern Australia and from a cultivated worldwide collection held in Sydney, Australia. Nat Toxins 2:347–353

Smith BL, Seawright AA, Ng JC, Hertle AT, Thomson JA, Bostock PD (1994b) Concentration of ptaquiloside, a major carcinogen in bracken fern (*Pteridium* spp.), from Eastern Australia and from a cultivated worldwide collection held in Sydney, Australia. Nat Toxins 2:347–353

Smith BL, Shaw G, Prakash AS, Seawright AA (1994c) Studies on DNA formation by ptaquiloside, the carcinogen of bracken ferns *Pteridium* spp. In: Colegate SM, Dorling PR (eds) Plant associated toxins: agricultural, phytochemical and ecological aspects. CAB International, Wallingford, UK, pp 167–172

Smith ND, Rubenstein JN, Eggener SE, Kozlowski JM (2003) The p53 tumour suppressor gene and nuclear protein: basic science review and relevance in the management of bladder cancer. J Urol 169:1219–1228

Somvanshi R, Lauren DR, Smith BL, Dawra RK, Sharma OP, Sharma VK, Singh AK, Gangwar NK (2006) Estimation of the fern toxin, ptaquiloside, in certain Indian ferns other than bracken. Curr Sci 91:1547–1552

Sood S, Dawra RK, Sharma OP, Kurade NP (2003) Exposure to the fern *Onychium contiguum* causes increase in lipid peroxidation and alters antioxidant status in urinary bladder. Biochem Biophys Res Commun 302:476–479

Ushijima J, Matsukawa K, Yuasa A et al (1983) Toxicities of bracken fern in guinea pigs. Jpn J Vet Sci 45:593–603

van der Hoeven JCM, Lagerweij WJ, Posthumus MA, Veldhuizen A, Holterman HAJ (1983) Aquilide A, a new mutagenic compound isolated from bracken fern (*P. aquilinum* (L.) Kuhn). Carcinogenesis 4:1587–1590

Villalobos-Salazar L, Hernandez H, Meneses A, Salazar G (1999) Bracken fern: toxicity, biology and control. In: Taylor JA, Smith RT (eds) Proceedings of the International Bracken Group, 4th International Bracken Conference, University of Manchester, pp 68–75

Wosiacki SR, Barreiro MAB, Alfieri AF, Alfieri AA (2005) Semi-nested PCR for detection and typing of bovine *Papillomavirus* type 2 in urinary bladder and whole blood from cattle with enzootic haematuria. J Virol Methods 126:215–219

Wosiacki SR, Claus MP, Alfieri AF, Alfieri AA (2006) Bovine papillomavirus type 2 detection in the urinary bladder of cattle with chronic enzootic haematuria. Mem Inst Oswaldo Cruz, Rio de Janeiro 101:635–638

Yoshida M, Saito T (1994) Acute toxicity of braxin C, a bracken toxin, in guinea pigs. J Toxicol Sci 19:17–23

Methods for Deriving Pesticide Aquatic Life Criteria for Sediments

Tessa L. Fojut, Martice E. Vasquez, Anita H. Poulsen, and Ronald S. Tjeerdema

Contents

1   Introduction 98
2   Summary of Major Sediment Quality Criteria Derivation Approaches 100
    2.1   Mechanistic Approach (Equilibrium Partitioning) 104
    2.2   Empirical Approaches 107
    2.3   Spiked-Sediment Toxicity Test Approach 108
3   Definitions and Uses of Sediment Quality Criteria 108
    3.1   Numeric Criteria Versus Advisory Concentrations 110
    3.2   Definitions and Uses of Numeric Criteria 111
4   Protection and Confidence 112
    4.1   Level of Biological Organization to Protect 112
    4.2   Portion of Species to Protect 113
    4.3   Probability of Over- and Under-protection 113
5   Data Required for Deriving Sediment Quality Criteria 114
    5.1   Data Sources and Literature Searches 114
    5.2   Physicochemical Data 117
    5.3   Ecotoxicity Data 120
    5.4   Quantitative Structure Activity Relationships (QSARs) 130
    5.5   Data Combination and Exclusion 131
6   Criteria Calculation 131
    6.1   Exposure Considerations 132
    6.2   Summary of Methodologies 139
    6.3   Important Additional Considerations for SQC Derivation 155
7   Summary 162
References 165

T.L. Fojut (✉) • M.E. Vasquez
Department of Environmental Toxicology, College of Agricultural and Environmental Sciences,
University of California, Davis, CA 95616-8588, USA

Central Valley Regional Water Quality Control Board, 11020 Sun Center Dr. Ste. 200,
Rancho Cordova, CA 95670-6114, USA
e-mail: tlfojut@ucdavis.edu

A.H. Poulsen • R.S. Tjeerdema
Department of Environmental Toxicology, College of Agricultural and Environmental Sciences,
University of California, Davis, CA 95616-8588, USA

D.M. Whitacre (ed.), *Reviews of Environmental Contamination and Toxicology*,   97
Reviews of Environmental Contamination and Toxicology 224,
DOI 10.1007/978-1-4614-5882-1_4, © Springer Science+Business Media New York 2013

1   Introduction

Sediments represent an integral component of aquatic ecosystems that provide habitat and food sources for aquatic life. Although many international and local governments have regulations in place to protect aquatic life, the majority of these are focused on or are based upon criteria for conserving water quality, and these have been in place for many years. After release to the aqueous environment, however, various organic chemicals tend to accumulate in sediments, where they may cause toxicity to aquatic life, even when water quality criteria (WQC) are met. Sediments comprise a complex medium and pose unique challenges for those who would like to develop single numeric concentrations below which aquatic life is protected. Currently, no official US Environmental Protection Agency (USEPA) methods or other agreed upon approaches are available in the United States for generating such sediment quality criteria (SQC).

The potential for a compound to accumulate in sediments or tissues or biomagnify up the food chain is determined by environmental factors and by the physicochemical properties of the compound. Chemical partitioning from water to sediment is usually driven by the chemical's insolubility in water; however, the mechanism by which chemicals sorb to sediments can also be influenced by pH, temperature, and sorbate properties—grain size, organic carbon (OC) content and makeup (e.g., black carbon), clay and mineral content, redox potential, and moisture content (Schwarzenbach et al. 2003). The availability of sediment-bound chemicals for uptake in organisms (i.e., bioavailability) and the related toxic effects are limited by the particular sorbent properties, chemical properties, and organism behavior at the site of interest (Day et al. 1995). Sediment composition varies greatly, both spatially and temporally, and some of this variability can be accounted for by normalizing contaminant concentrations for different sediment characteristics such as OC and fine matter fractions. Indeed, many studies have demonstrated that total measured contaminant concentrations in sediments are poorly correlated to observed effects, and this is most likely due to limited contaminant bioavailability (Conrad et al. 1999; Di Toro et al. 2002; Xu et al. 2007). Benthic organisms further introduce multiple exposure routes (i.e., diffusion from both water and sediment, ingestion of sediment), which must be characterized to comprehensively assess exposure. The importance of characterizing the factors that determine chemical partitioning and bioavailability within the sediment compartment is a crucial point of difference between developing criteria for sediment versus water quality criteria.

The state of California has nine Regional Water Quality Control Boards (RWQCBs) whose mission is "to develop and enforce water quality objectives and implementation plans, which will best protect the beneficial uses of the State's waters, recognizing local differences in climate, topography, geology and hydrology" (California SWRCB 2011). Toward that mission, each RWQCB must develop a "basin plan" for its specific hydrologic area. The "Water Quality Control Plan (Basin Plan) for the Sacramento River and San Joaquin River Basins" aspires to maintain specific goals for toxic substances in general, and pesticides in particular, and states that "…waters shall be maintained free of toxic substances in concentrations that produce detrimental physiological responses in human, plant, animal, or aquatic life," "no individual pesticide or combinations of pesticides shall be present in concentrations that adversely affect beneficial uses," "discharges shall not result in pesticide concentrations in bottom sediments or aquatic life that adversely affect beneficial uses," and "pesticide concentrations shall not exceed the lowest levels technically and economically achievable" (CRWQCB-CVR 2009). A recent paper reviewed methodologies for deriving WQC for pesticides with particular reference to agents relevant for the Sacramento and San Joaquin River basins (TenBrook et al. 2009). In the current paper, we review existing and proposed SQC derivation methodologies from around the world as part of a larger project that aims to develop a methodology for deriving pesticide SQC for the protection of aquatic life in these Californian river basins.

The surface waters of the Sacramento and San Joaquin River basins receive pesticide inputs in runoff and drainage from agriculture, silviculture, and residential and industrial storm water (CRWQCB-CVR 2009). Pesticides most likely to cause sediment toxicity are nonpolar nonionic organic compounds, which tend to sorb to solids and colloids in aqueous environments. Nonpolar pesticides have been detected in California freshwater bedded and/or suspended sediments in recent years. Those detected include herbicides (dimethyl tetrachloroterephthalate [chlorthal-dimethyl or DCPA], ethalfluralin, metolachlor, oxyfluorfen, pendimethalin, prometryn, simazine, trifluralin), organochlorine insecticides (DDT and metabolites, dieldrin, endosulfan, endrin, lindane, methoxychlor), organophosphate insecticides (chlorpyrifos, diazinon, methylparathion), and pyrethroid insecticides (bifenthrin, cyfluthrin, cypermethrin, deltamethrin, esfenvalerate, fenpropathrin, λ-cyhalothrin, permethrin; Domalgalski et al. 2010; Hladik and Kuivila 2009; Holmes 2004; Weston et al. 2004, 2005). Most of the organochlorines that have been recently detected are no longer used; in contrast, use of pyrethroid insecticides has increased over the last decade, as they are seen as replacements for the organophosphates. Pyrethroids are used in both agricultural and urban/residential settings and are characterized by extreme insolubility in water with high degrees of sorption to solids, including soils, sediment, and dissolved organic matter (Laskowski 2002). Although pyrethroids have low mammalian toxicity, they are highly toxic to aquatic invertebrates and fish, and their use has not been without adverse effects. Pyrethroids have been detected in sediments of both urban and agricultural waterways at levels that are toxic to the freshwater invertebrate *Hyalella azteca* in the laboratory (e.g., 10-day median lethal concentrations (LC_{50}) ranging from 0.38 to 10.83 µg/g OC for cypermethrin and permethrin, respectively; Amweg et al. 2006; Weston et al. 2004). In this review, we focus on methodologies that are suitable for setting pesticide criteria, with particular emphasis on the pyrethroids.

Several scientific documents are available in which current methodologies are evaluated for SQC derivation. These documents include two publications relating to a Society of Environmental Toxicology and Chemistry (SETAC) Pellston workshop entitled "Use of Sediment Quality Guidelines and Related Tools for the Assessment of Contaminated Sediments" (Wenning and Ingersoll 2002; Wenning et al. 2005), a report entitled "Review and Recommendations of Methodologies for the Derivation of Sediment Quality Guidelines" (Rowlatt et al. 2002), a review paper by Chapman (1989) entitled "Current Approaches to Developing Sediment Quality Criteria," and a report prepared on behalf of the European Commission (EC) entitled "Towards the Derivation of Quality Standards for Priority Substances in the Context of the Water Framework Directive" (Lepper 2002). These previous review documents drew attention to a range of aspects important for evaluation and development of a SQC derivation methodology. The relevant components identified in these reports are outlined in Table 1 and form the backbone of the current review. For each component, a brief introduction is presented to describe its relevance for SQC derivation, and thereafter, international methodologies are reviewed for their incorporation or disregard of the component. The reviewed methodologies include those of Australia/New Zealand, Canada, the European Union/European Commission (EU/EC), France, the Netherlands, the Organisation for Economic Co-operation and Development (OECD), the United Kingdom (UK), and the United States (USA) including those of the USEPA and the National Oceanic and Atmospheric Administration (NOAA), as well as a few state regulations. Where original documents were not available in English, other resources containing summaries of those documents were used, if available.

In the present review, we focus on methodologies suitable for developing numeric SQC for which compliance can be based solely on chemistry measurements. Most current methodologies refer to sediment quality guidelines (SQGs), which are recommended for use as part of a risk assessment framework or as triggers for further research rather than enforceable criteria (i.e., SQC). Most risk assessment literature focuses on metals and industrial and legacy chemicals, whereas pesticides are the specific objective of this review. Methodologies that primarily focus on metals, dredged materials, and marine environments have been excluded from or are not the focus of this review, because freshwater environments with pesticide contamination pose significantly different issues than metal-contaminated harbors. Available data for current-use hydrophobic pesticides, such as the pyrethroids, are relatively sparse. Supplementary information is therefore included for some industrial chemicals and legacy pesticides (i.e., PCBs, PAHs, DDT and metabolites, and organochlorines) that exhibit similar physicochemical properties as current-use hydrophobic pesticides.

In Table 2, we define the many acronyms used throughout this review.

2 Summary of Major Sediment Quality Criteria Derivation Approaches

Current methodologies employ many different types of numeric sediment quality criteria depending on their intended use and the quantity and quality of available data. Table 3 outlines the principles and criteria types of the most commonly employed methodologies for deriving SQC. Our goal in this review is to evaluate the suitability of any existing methodology or combination of methodologies for

Table 1 Relevant components for deriving sediment quality criteria

Category	Component	References
Criteria types (Sect. 3)	Numeric criteria vs. advisory concentrations	a, b
	Multiple levels of criteria	c, d
Protection level (Sect. 4)	Protect all species to protect ecosystem	b, c, e
	Protect aquatic environment	d, f, g, h, i, j, k
	Benthic communities	l
	Probability of over or under protection	c
Physicochemical and ecotoxicological data (Sect. 5)	Data sources	c, m, n
	Literature search protocol	b, c, f, k
	Physicochemical data requirements and quality	c, e, f, h, j
	Acute vs. chronic sediment toxicity testing	b, c, o
	Laboratory vs. field data	c, k
	Traditional vs. nontraditional endpoints	b, c, k
	Ecotoxicity data quality	b, c, o
	Ecotoxicity data quantity	b, p
	Quantitative structure activity relationships (QSARs)	c
Criteria calculation (Sect. 6)	Magnitude, duration, and frequency of exposure	b, e
	Multiple exposure routes	k
	Bioavailability	c, e, f, j, k
	Equilibrium partitioning	c, d, e, f, h, j, k
	Suspended sediments vs. bedded sediments	d, k
	Spiked-sediment toxicity testing approach	b, c, k
	Standardized sediment	c, f
	Assessment factors	b, c, d, k
	Species sensitivity distribution	c, k
	Mixtures	k
	Bioaccumulation and secondary poisoning	c, p
	Encouragement of data generation	b, c, d, k
	Utilization of available data	c, k
	Harmonization (equilibrium partitioning, EqP)	c, d, e, f, h, j, k

[a]USEPA 2003a, b, c
[b]CCME 1995
[c]RIVM 2001
[d]Lepper 2002
[e]Di Toro et al. 2002
[f]Persaud et al. 1993
[g]MacDonald 1994
[h]Rowlatt et al. 2002
[i]ANZECC and ARMCANZ 2000
[j]OECD 1995
[k]ECB 2003
[l]Diaz and Rosenberg 1996
[m]TenBrook et al. 2009
[n]USEPA 1985
[o]Cubbage et al. 1997
[p]SWRCB 2011

Table 2 List of acronyms and abbreviations used in this review

AET	Apparent effects threshold
AF	Assessment factor
ANZECC	Australia and New Zealand Environment and Conservation Council
ARMCANZ	Agriculture and Resource Management Council of Australia and New Zealand
ASTM	American Society for Testing and Materials
BAF	Bioaccumulation factor
BCF	Bioconcentration factor
BMF	Biomagnification factor
BEDS	Biological effects database for sediments
BSAF	Biota-sediment accumulation factor
CAS	Chemical abstracts service
CCME	Canadian Council of Ministers of the Environment
CDFG	California Department of Fish and Game
CDPR	California Department of Pesticide Regulation
C_{iw}	Chemical concentration in interstitial water
CRWQCB-CVR	California Regional Water Quality Control Board, Central Valley Region
$C_{s,oc}$	Chemical concentration in whole sediment or organic carbon
Ctgb	Dutch Board for the Authorization of Plant Protection Products and Biocides
DCPA	Dimethyl 2,3,5,6-tetrachlorobenzene-1,4 dicarboxylate (or dimethyl tetrachloroterephthalate)
DDT	Dichlorodiphenyltrichloroethane
DOC	Dissolved organic carbon
DOM	Dissolved organic matter
DT_x	Time for $x\%$ of a chemical to degrade
EC	European Commission
EC_x	Concentration that affects x% of exposed organisms
ECB	European Chemicals Bureau
EPA	Environmental Protection Agency
ERL	Effects range low
$ERL(sed_{EP})$	Environmental risk limit for the sediment compartment using EqP theory
ERL(water)	Environmental risk limit for aquatic species
ERM	Effects range median
EINECS	European Inventory of Existing Commercial Substances
EqP	Equilibrium partitioning
ESG	Equilibrium partitioning sediment guideline
EU	European Union
FCV	Final chronic value
f_{oc}	Fraction of organic carbon
H	% organic matter content
HC_p	Hazardous concentration potentially harmful to $p\%$ of species
HOC	Hydrophobic organic compound
HPLC	High performance liquid chromatography
HSDB	Hazardous Substance Data Bank
IUPAC	International Union of Pure and Applied Chemistry
ISQG	Interim sediment quality guideline
IWQ	Interstitial water quality

(continued)

Table 2 (continued)

K_d	Solid-water partition coefficient
K_{oc}	Organic carbon-normalized solid-water partition coefficient
K_{ow}	Octanol-water partition coefficient
K_p	Solid-water partition coefficient (equivalent to K_d)
$K_{s/l}$	Sediment-liquid partition coefficient
$K_{susp-water}$	Suspended matter-water partition coefficient
K_x	Interaction coefficient for a synergist/antagonist at concentration x
LC_x	Concentration lethal to $x\%$ of exposed organisms
LEL	Lowest effect level
LOEC	Lowest observed effect concentration
LOEL	Lowest observed effect level
MATC	Maximum acceptable toxicant concentration
MPC	Maximum permissible concentration
MTC	Maximum tolerable concentration
NC	Negligible concentration
NEL	No effect level
NOAA	National Oceanic and Atmospheric Administration
NOEC	No observed effect concentration
NOEL	No observed effect level
NSTP	National Status and Trends Program
NTIS	National Technical Information Service
OC	Organic carbon
OCSPP	Office of Chemical Safety and Pollution Prevention
OECD	Organisation for Economic Co-operation and Development
OM	Organic matter
OMEE	Ontario Ministry of Environment and Energy
OPP	Office of Pesticide Programs
OPPTS	Office of Prevention, Pesticides and Toxic Substances
OPPT	Office of Pollution Prevention and Toxics
PAET	Probable apparent effects threshold
PAH	Polycyclic aromatic hydrocarbons
PCB	Polychlorinated biphenyls
PEC	Probable effect concentration
$PEC_{oral, predator}$	Predicted chemical concentration a predator will receive in prey (food)
PEC_{sed}	Measured or predicted chemical concentration in sediment
PEC_{water}	Measured or predicted chemical concentration in water
PEL	Probable effect level
pK_a	Acid dissociation constant
PNEC	Predicted no effect concentration
$PNEC_{sed}$	Predicted no effect concentration in sediment
$PNEC_{water}$	Predicted no effect concentration in water
PWQO/G	Provincial Water Quality Objectives/Guidelines
QSAR	Quantitative structure activity relationship
RHO_{susp}	Bulk density of wet suspended matter
RIVM	National Institute for Public Health and the Environment, Bilthoven, The Netherlands

(continued)

Table 2 (continued)

RL	Risk limit
RWQCB	Regional Water Quality Control Board
SEL	Severe effect level
SETAC	Society of Environmental Toxicology and Chemistry
SLC	Screening level concentration
SPME	Solid-phase microextraction
SRC_{ECO}	Ecosystem serious risk concentration
SSD	Species sensitivity distribution
SSLC	Species screening level concentration
SSTT	Spiked-sediment toxicity testing
SQC	Sediment quality criteria
SQG	Sediment quality guideline
SQG_{oc}	OC-normalized sediment quality guideline
SWRCB	State Water Resources Control Board
TEC	Threshold effect concentration
TEL	Threshold effect level
TES	Threatened and endangered species
TU	Toxic unit
UK	United Kingdom
US	United States
USEPA	US Environmental Protection Agency
WQC	Water quality criterion/criteria

calculating pesticide SQC. Three main approaches are currently available to develop sediment quality guidelines: empirical, mechanistic, and spiked-sediment toxicity testing (SSTT). In general, the empirical approaches generate concentration ranges that are very likely, likely, or not likely to cause adverse effects, while the mechanistic approaches generate single recommended maximum concentrations that are based on existing water quality criteria for the compound of interest. The third approach applies spiked-sediment toxicity data to derive SQC. Depending on data availability, the SSTT approach allows quantification of statistical uncertainty in the derived criteria. Several current methodologies incorporate multiple approaches and recommend the SSTT approach when sufficient data are available. The three major approaches for SQG derivation are introduced briefly in Sects. 2.1 through 2.3.

2.1 Mechanistic Approach (Equilibrium Partitioning)

The partitioning of a chemical between a solid and water can be described via the solid-water partition coefficient (K_d), which is determined as the chemical concentration in the solid divided by the chemical concentration in water after chemical equilibrium has been reached between the two phases (Schwarzenbach et al. 2003). Because sediment composition varies greatly over space and time, sediment-contaminant concentrations may be normalized to different sediment properties for

Table 3 Summary of major methodologies

Method title	Source	Jurisdiction	Criterion	Criterion derivation
Mechanistic approaches—equilibrium-partitioning (EqP) model				
Technical basis for the derivation of equilibrium-partitioning sediment quality guidelines for the protection of benthic organisms: nonionic organics	USEPA (Di Toro et al. 2002)	US	Tier 1 ESG: equilibrium-partitioning sediment guideline	ESG is derived using water quality criterion (WQC) and solid-water partition coefficient K_d (or K_{oc}). Compare with SSTT data; tiered approach depends on data availability
			Tier 2 ESG	If SSTT data are limited/unavailable to confirm EqP prediction
Guidance document on deriving environmental risk limits	RIVM (2001)	Holland	MPC: maximum permissible concentration	Sediment MPC is derived from WQC and K_d (or K_{oc}) and compared with SSTT data
Guidance document for aquatic effects assessment	OECD (1995)	OECD	MTC$_{sed}$: maximum tolerable concentration for sediment	MTC$_{sed}$ is derived from WQC and K_d (or K_{oc})
Recommendations on the development of sediment quality guidelines	UK (Rowlatt e al. 2002)	UK	SAL: sediment action level	SAL is derived from WQC and K_{oc}
Guidelines for the protection and management of aquatic sediment quality in Ontario	OMEE (Persaud et al. 1993)	Ontario	SQG: sediment quality guideline	SQG is derived from WQC and K_{oc}
Spiked-sediment toxicity testing (SSTT) methodologies				
Technical guidance document on risk assessment	EU (ECB 2003)	EU	PNEC: predicted no effect concentration	For adequate SSTT data, use SSD approach. For limited SSTT data, apply assessment factor (AF) (acute or chronic). If no data, use EqP
Towards the derivation of quality standards for priority substances in the context of the water framework directive	EU (Lepper 2002)	France	Threshold level 1 Threshold level 2	AF approach: lowest NOEC/10 or L(E)C$_{50}$/1,000 AF approach: lowest NOEC or L(E)C$_{50}$/100

(continued)

Table 3 (continued)

Method title	Source	Jurisdiction	Criterion	Criterion derivation
Empirical approaches				
Sediment quality guidelines developed for the National Status and Trends Program (NSTP)	NOAA, NSTP (Long and Morgan 1990)	US	ERL/ERM: effects range low/median	Utilizes large matching sediment chemistry and biological effects database; ERL and ERM defined as concentration at 10th and 50th percentile of NSTP database, respectively
Australian and New Zealand guidelines for fresh and marine water quality	ANZECC and ARMCANZ (2000)	Australia/ New Zealand	ERL/ERM	Based on North American database refined with local sediment data if available
Combination of methodologies (empirical and SSTT)				
Protocol for the derivation of Canadian sediment quality guidelines for the protection of aquatic life	CCME (1995)	Canada	ISQG: interim sediment quality guideline	Empirical approach based on NSTP database for derivation of interim sediment quality guidelines
			SQG: sediment quality guideline	Use SSTT approach to establish cause/ effect relationships between chemical concentration and response

ANZECC = Australia and New Zealand Environment and Conservation Council, ARMCANZ = Agriculture and Resource Management Council of Australia and New Zealand, CCME = Canadian Council of Ministers of the Environment, ECB = European Chemicals Bureau, OECD = Organisation for Economic Co-operation and Development, RIVM = National Institute for Public Health and the Environment, the Netherlands, SSD = species sensitivity distribution the purpose of reducing variability. For example, organic compounds are primarily thought to sorb to organic carbon in sediments and are thus normalized to OC content to reduce variability in sorption and toxicity measures, although some variation often remains (Amweg et al. 2005; Xu et al. 2007). Dry weight normalization has been used to achieve similar or better success for hydrophobic organic compounds (HOCs), such as PCBs and PAHs (Ingersoll et al. 2000). Fine particles have relatively high surface areas, and they tend to be enriched in organic contaminants compared to coarser sediments; thus, normalization for fine matter content has also been proposed (Mudroch and Azcue 1995). Concentrations are commonly normalized to OC, and K_d is often expressed as K_{oc}, which is defined as K_d divided by the sediment OC fraction (Schwarzenbach et al. 2003).

In the 1990s, the USEPA developed a mechanistic approach for generating sediment quality guidelines, based on total chemical concentrations using the equilibrium-partitioning (EqP) model (Di Toro et al. 2002; USEPA 1993). In this approach, it is assumed that (1) toxicity is only caused by the freely dissolved fraction of a contaminant, (2) this fraction is in equilibrium—between sediment and porewater (interstitial water between sediment particles)—and (3) chemical exposure is equivalent in each of these environmental compartments. The EqP approach applies the final chronic value (FCV) from a chemical's water quality criterion (WQC) and K_d (or K_{oc}) to derive an equilibrium-partitioning sediment guideline (ESG; Di Toro et al. 2002; USEPA 1993):

$$ESG = FCV \times K_d (or \ K_{oc})$$ (1)

The FCV is equivalent to the chronic criterion, or criterion continuous concentration (CCC), as derived by the USEPA water quality criteria derivation methodology (USEPA 1985). The FCV is the concentration of a chemical intended to be protective of aquatic life over chronic exposure. To justify using the water-based FCV for deriving an ESG, the EqP method assumes that epibenthic and benthic organisms have the same species sensitivity distribution as water column organisms (Di Toro et al. 2002).

Organic carbon is assumed to be the primary factor governing the partitioning of nonionic organic chemicals between sediment and porewater. EqP applications for such compounds, including PAHs, PAH mixtures, and several organochlorine pesticides, have therefore typically used K_{oc} as the partition coefficient. The EqP approach is employed by many of the methodologies reviewed, including those of the Netherlands (Kalf et al. 1999; RIVM 2001), the EU (ECB 2003), Ontario (Persaud et al. 1993), France (Lepper 2002), and the OECD (1995). This mechanistic approach can also be used to establish SQGs with other lines of evidence, such as comparison to field concentrations or spiked-sediment toxicity data.

2.2 Empirical Approaches

Empirical approaches utilize databases for field-collected sediments comprising both sediment chemistry and observed biological effects data, which are correlated to determine numerical chemical ranges for various effects levels. The NOAA empirical method for deriving SQGs was initially developed to interpret chemical data from a comprehensive monitoring program named the National Status and Trends Program (NSTP; Long and Morgan 1990). The NSTP SQGs were not intended for regulatory purposes, but for informal use such as ranking areas for further study. Empirical approaches demonstrate correlations between observed effects and chemical concentrations, but do not elucidate causative relationships for the observed toxicity. The observed effects may be caused by the chemicals measured but could also be caused by other nontarget chemicals or by environmental conditions. Sediment toxicity is often a result of mixture effects. Chemical analysis and toxicity testing of field sediments do not provide the information necessary to attribute toxicity to any individual chemical, and as such, empirical approaches are inappropriate for deriving criteria for individual chemicals. The limitations of empirical approaches are well known, and recent recommendations have disclosed that this type of data must be considered with other biological effects information in a multiple lines of evidence approach (Wenning et al. 2005). The NOAA NSTP (Long and Morgan 1990) empirical methodology is widely accepted and has been adapted by many other jurisdictions including Australia/New Zealand (ANZECC and ARMCANZ 2000), Canada (CCME 1995), Ontario (Persaud et al. 1993), UK (Rowlatt et al. 2002), Washington State (Cubbage et al. 1997), California (California SWRCB 2011), and Florida (MacDonald 1994).

2.3   Spiked-Sediment Toxicity Test Approach

The spiked-sediment toxicity test (SSTT) approach is similar to methods used for WQC derivation. Acute and chronic toxicity data from controlled, spiked-sediment laboratory experiments are used to set SQGs. If data are abundant, a species sensitivity distribution (SSD) can be used to derive an appropriate effect level that accounts for statistical uncertainty. If data are sparse, assessment (or safety) factors may be applied to the lowest concentration lethal to 50% of exposed organisms (LC_{50}) or the lowest NOEC. Assessment factors can also be applied to account for uncertainty from various other factors or from data scarcity. Methodologies that include a SSTT approach are those from Canada (CCME 1995), the Netherlands (RIVM 2001), the EU (ECB 2003), and France (Lepper 2002).

3   Definitions and Uses of Sediment Quality Criteria

The USEPA is authorized to develop and implement sediment quality criteria under Section 304(a) of the Clean Water Act (2002). The USEPA has recommended that states use numeric criteria with sediment bioassays to interpret the narrative criteria, which are typically stated as "no toxics in toxic amounts," (USEPA 1998).

The Central Valley Regional Water Quality Control Board can use numeric SQC to set sediment quality objectives or total maximum daily loads. A numeric criterion is a chemical threshold concentration that has been scientifically derived to protect aquatic life from harmful effects of exposure to that chemical, without consideration of defined water body uses, societal values, or economics (TenBrook et al. 2009). This overall definition of numeric criteria applies for all methodologies, although criteria and guidelines are referred to by different terms by different jurisdictions. The terminology applied for numeric criteria includes the following: equilibrium-partitioning sediment guidelines (ESGs; Di Toro et al. 2002); effects range low and effects range median (ERL or ERM; Long and Morgan 1990); threshold effects concentrations and probable effects concentrations (TECs and PECs; MacDonald et al. 2000); no-effect level, lowest effects level, or severe effects level (NEL, LEL, or SEL; Persaud et al. 1993); maximum permissible concentrations (MPC; Kalf et al. 1999); and predicted no-effect concentrations (PNECs; ECB 2003).

Chapman (1989) stated that SQC development is important "because (a) various toxic contaminants found in only trace amounts in the water column accumulate in sediments to elevated levels, (b) sediments serve as both a reservoir and a source of contaminants to the water column, (c) sediments integrate contaminant concentrations over time, whereas water column contaminant concentrations are much more variable, (d) both sediment contaminants and water column contaminants affect benthic and other sediment-associated organisms, and (e) sediments are an integral part of the aquatic environment, providing habitat, feeding and rearing areas for many aquatic biota."

SQC can refer to chemical-specific concentrations in sediment above which adverse biological effects are expected or to levels of biological effects that are considered unacceptable. In this review, we focus on SQC methods that yield chemical-specific numerical concentrations derived to be protective of aquatic life. The advantages to such criteria are their wide applicability and simplicity and that they do not require specialized biological, chemical, or other expertise (Chapman 1989). The disadvantages of individual numerical criteria are the risk of overlooking mixture toxicity and the inability to account for site-specific variations, particularly the bioavailability and subsequent toxicity of sediment-bound chemicals.

Numeric SQC can be used in a similar manner as numeric WQC are used (USEPA 1993). One difference between WQC and SQC is that the regulatory basis and implementation of SQC are yet to be established, whereas WQC have been implemented to regulate water contaminants, since many were derived using the 1980 USEPA guidelines (USEPA 1980). The application of SQC may further differ from WQC in that water column contaminants can often be controlled by limiting the sources, whereas toxicity caused by previously accumulated sediment contaminants may continue despite source limitation (USEPA 1993). However, source control could be effective for current-use pesticides, particularly those with short to moderate half-lives that will likely dissipate over a short timescale compared to legacy contaminants.

Many researchers have cautioned against the use of numeric SQC for regulatory purposes or as pass/fail criteria. As noted, many methodologies thus refer to the derived values as guidelines rather than criteria; they also recommend using these SQGs as part of a risk assessment framework or as triggers for further research, rather than as enforceable criteria. Burton (2002) concluded, for example, that due to the complexity of sediments and mixture interactions, SQGs will probably always be used as screening tools rather than as enforceable regulatory values. Burton further explains that SQGs do not characterize microscale variation, inorganic speciation differences, stressor interactions, dynamics of biota, or critical physicochemical parameters (Burton 2002). Chapman is also wary of the emphasis placed on numeric chemical values and points out that while sediment chemistry can help identify areas, sources, and contaminants of concern, such data do not provide any information on bioavailability or toxicity (Chapman 2000, 2007). Chapman and Mann (1999) identified the key limitations of numeric SQGs as being the following: (1) degree of conservatism, inaccurate and uncertain numeric results tend toward under- or overprotection; (2) bioaccumulation/biomagnification, SQGs are usually derived from direct toxicity data that do not address effects of long-term bioaccumulation and biomagnification; (3) bioavailability, SQGs cannot be applied to all sediment conditions because bioavailability varies greatly depending on conditions (although mechanistic approaches attempt to normalize for this variation); (4) contaminant mixtures, SQGs are often based on data from field samples that likely contain chemical mixtures, which can confound SQG databases; and (5) predictability, the accuracy of SQG predictions of adverse ecological effects is debatable as high percentages of false-positives and/or false-negatives have been observed.

3.1 Numeric Criteria Versus Advisory Concentrations

The USEPA has derived numeric SQC for dieldrin, endrin, and 34 PAHs (USEPA 2003a, b, c), but these have not been adopted as enforceable sediment quality standards. Instead, numeric criteria may be used as guidelines for interpreting site-specific sediment chemistry data as part of an environmental risk assessment framework (i.e., predicting the degree and extent of contamination) or possibly to implement narrative criteria. The guidelines are considered to be advisory concentrations because current evidence does not support numeric SQC as conclusive predictors of effects. The USEPA divides the advisory concentrations into Tier 1 and Tier 2 categories, in which Tier 1 guidelines are calculated with more data and are thus associated with higher certainty than Tier 2 guidelines (Di Toro et al. 2002). The Canadian method allows development of interim sediment quality guidelines when data are limited, yet both interim and full guidelines may be used as a basis to set enforceable, site-specific sediment quality objectives (CCME 1995).

   Numeric SQGs are used in many risk assessment frameworks in the early assessment tiers to identify potentially toxic contaminant levels (NOAA, Washington State, Great Lakes; USEPA 1994a). When a contaminant exceeds a SQG in risk assessment, this is typically a trigger for further investigation, but not a basis for regulatory action. Further investigation in the higher tiers of risk assessment usually include field sediment bioassays to assess if adverse effects are likely caused by sediment toxicity, and these results can lead to various management decisions (Apitz and Power 2002).

3.2   Definitions and Uses of Numeric Criteria

Many water quality criteria derivation methodologies include procedures for deriving more than one level or type of criterion for each toxicant (ANZECC and ARMCANZ 2000; La Point et al. 2003; Lepper 2002; OECD 1995; RIVM 2001; USEPA 2003d). For deriving sediment quality criteria, the available sediment toxicity data or knowledge are typically insufficient to enable derivation of different criteria levels to meet different regulatory goals (i.e., enforcement vs. risk assessment). The methodologies of the Netherlands and France, however, do offer estimation of more than one type of sediment criteria.

In the Netherlands, compartment-specific environmental risk limits (RLs) for water and sediment are derived by the same general protocol, which includes three levels, the ecosystem serious risk concentration (SRC_{ECO}), the maximum permissible concentration (MPC), and the negligible concentration (NC) (RIVM 2001). The MPC is intended to protect all species in an ecosystem from adverse effects. The NC is derived by dividing the MPC by a safety factor of 100 and is used as a regulatory target value at which ecosystems are expected to experience negligible effects. If sediment-contaminant concentrations exceed the MPC, discharges can be further regulated. The SRC_{ECO} represents concentrations that will cause ecosystem functions to be seriously affected or threatened to be negatively affected (i.e., when 50% of species and/or 50% of microbial and enzymatic processes are possibly affected; RIVM 2001). Sediments contaminated in excess of the SRC_{ECO} require cleanup intervention efforts.

The French protocol characterizes four threshold levels (TLs) for sediment and suspended matter contaminants that exhibit log-normalized octanol-water partition coefficients (log K_{ow}) >3 (Lepper 2002). Each TL corresponds to a different biological quality suitability class for water bodies. TL1 represents negligible risk for all species and is derived from chronic or acute toxicity data with assessment (or safety) factors applied. TL2 represents concentrations causing possible risk of adverse chronic sublethal effects for the most sensitive species and is derived in the same manner as TL1 but uses smaller assessment factors. Threshold levels 3 and 4 indicate a probable and a significant risk of adverse ecosystem effects, respectively, but have never been calculated for sediment due to a lack of required data. Because of the inherent uncertainties in the two methods used to calculate threshold levels for sediment and suspended particles (i.e., equilibrium partitioning and weight of evidence), these values are all considered provisional. The threshold values derived by the French methodology are not enforceable but serve as references for risk assessment and actions.

Regardless of terminology, all criteria and guidelines discussed were developed with the common goal of estimating concentrations of chemicals that, if exceeded, may lead to loss of designated uses of water bodies. The statistical certainty and site specificity of the derived values will be determined by the quantity and quality of available data for the site and compound of interest.

4 Protection and Confidence

Aquatic life sediment quality criteria aim to protect aquatic life from exposure to toxic substances. The protection of aquatic life can be defined in various ways, from overall ecosystem protection to protection of each individual in the ecosystem. The existing methodologies specify different protection goals in terms of the level of ecosystem organization and how to approximate the protection level by extrapolating ecosystem effects from existing data. The ability to state with a quantified level of certainty that criteria achieve the intended level of protection is also important (TenBrook et al. 2009).

4.1 Level of Biological Organization to Protect

A comprehensive description of how the levels of ecosystem organization are defined can be found in TenBrook et al. (2009). Most of the methodologies reviewed designate what level of organization is to be protected by SQC. Several protocols for SQC derivation seek to protect each species of an ecosystem for the purpose of the entire ecosystem. The aim of the Canadian guiding principles for the development of numerical SQGs is that these are "set with the intention to protect all forms of aquatic life and all aspects of the aquatic life cycle" (CCME 1995). The goal of the USEPA is to be protective of benthic aquatic species, as stated in their EqP methodology (Di Toro et al. 2002). The Netherlands also aims to protect all species in ecosystems from adverse effects (RIVM 2001).

It is the ambition of most of the reviewed methodologies to protect aquatic ecosystems. France targets threshold levels intended to maintain an ecosystem's suitability to support its biological function and other uses (Lepper 2002). The province of Ontario, Canada, states that the purpose of SQGs is to protect the aquatic environment (Persaud et al. 1993). The state of Florida has the goal of protecting living resources and their habitats (MacDonald 1994). The objective for setting sediment action levels in the UK is the "maintenance of environmental quality so as to protect aquatic life and dependent nonaquatic organisms" (Rowlatt et al. 2002). Australia and New Zealand aspire "to maintain and enhance the "ecological integrity" of freshwater and marine ecosystems, including biological diversity, relative abundance and ecological processes" (ANZECC and ARMCANZ 2000). The OECD guidelines were developed for derivation of criteria "where no adverse effects on the aquatic ecosystem are expected" (OECD 1995). The predicted no-effect concentrations derived by the EU risk assessment methodology are aimed at ensuring "overall environmental protection" (ECB 2003). Finally, as discussed in the

Introduction (Sect. 1), the Basin Plan of the Central Valley Regional Water Quality Control Board states that "discharges shall not result in pesticide concentrations in bottom sediments or aquatic life that adversely affect beneficial uses" (CRWQCB-CVR 2009).

Diaz and Rosenberg (1996) point out that thus far, the functional component of benthic communities has been overlooked and that "at the ecosystem level, alteration or loss of function is of greatest concern." Benthic organisms mediate the cycling of materials between sediments and the overlying water column via burrowing, irrigation, and other behaviors. They comment that large, long-lived species are important for the vertical mixing of sediment to deep layers and for stabilizing annual productivity (Diaz and Rosenberg 1996). The elimination of such species thus affects key community functions that regulate mixing and energy flow, and the authors conclude that this is the main effect of sediment toxicants. When these larger species are reduced or eliminated, the food chains are shortened and there is less bioturbation, which leads to a higher tendency for contaminant accumulation in deep sediments (Diaz and Rosenberg 1996). Finally, these authors posit that functional changes are more indicative of ecosystem impacts than changes in community structure.

4.2 Portion of Species to Protect

In contrast to water quality criteria derivation methodologies, most SQC derivation methodologies do not primarily rely on single-species sediment toxicity data to calculate criteria or guidelines. Not only is there a dearth of spiked-sediment toxicity data, but also a lack of consistency in the data due to variable bioavailability across different sediments. Multispecies or ecosystem data are alternative options for use in criteria derivation, but these types of data are also relatively few and difficult to interpret. As summarized by TenBrook et al. (2009), protection of less than 100% of species may cause unpredicted harm to an ecosystem because each species performs a function and takes part in complex trophic interactions in the ecosystem structure. To ensure protection of entire ecosystems, both functions and structure must be maintained. The current review considers alternative extrapolation methods for estimating ecosystem no-effect concentrations from available toxicity data, which may be derived from aqueous toxicity data in the equilibrium-partitioning methods, from spiked-sediment toxicity data or from co-occurrence field data. Yet, the only way to confirm that SQC are truly protective of ecosystems is to perform field or semi-field studies.

4.3 Probability of Over- and Under-protection

For environmental managers to assess the ability of a criterion to provide the intended level of protection, such criterion is best expressed with associated confidence limits (TenBrook et al. 2009). Overprotective criteria may lead to unnecessary costs, whereas underprotective criteria risk adverse effects to the ecosystem of concern. The effect levels derived by the French method do not provide confidence limits because these threshold values are calculated by applying an assessment (or safety) factor to the single most sensitive datum (Lepper 2002). The EU also recommends applying an assessment factor to spiked-sediment toxicity test data if it is available (ECB 2003). These types of criteria may be protective, but do not allow evaluation of the degree to which the criteria are likely to over- or underprotect (TenBrook et al. 2009). Uncertainty analyses are possible with the EqP approach if sediment toxicity data are available for the compound of interest. Confidence limits can be estimated as the degree to which the sediment toxicity data are predicted by the EqP model (USEPA 2003b). The EU proposes that species sensitivity distributions can be used if data availability allows (ECB 2003). Since sediment data are generally lacking, however, the EU protocol does not provide full guidance on the use of SSDs with sediment data or on calculation of confidence limits. The Netherlands (RIVM 2001) employs a SSD technique that derives criteria at given confidence levels, but if ample data are unavailable, an assessment factor is applied to selected data, in which case this approach does not provide confidence limits. If a SSD is possible, confidence limits provide useful information. For a criterion derived with 50% confidence level, for example, the true no-effect level is equally likely to be above or below the derived criterion. If a criterion is derived at a 95% confidence level, on the other hand, only a 5% chance exists that the true no-effect level falls below the derived criterion. Such information on statistical reliability is valuable to environmental managers responsible for enforcing or recommending the criteria.

5 Data Required for Deriving Sediment Quality Criteria

Quality data are crucial for deriving scientifically sound SQC. All methodologies covered in this review require both physicochemical and ecotoxicity data. The quantity of data required to derive the criteria must also be adequate to minimize uncertainty. Ideally long-term (chronic) spiked-sediment toxicity test data for benthic organisms (considering various routes of exposure) should be available for contaminants across a wide range of sediment types. This information would allow criteria to be based on known cause and effect relationships. Although such data are limited, the goal of the following sections is to identify the quantity and quality of data required to generate numerical sediment criteria using various approaches worldwide.

5.1 Data Sources and Literature Searches

To evaluate, develop, and implement a methodology for SQC derivation, it is important to identify the data sources employed by the various agencies worldwide.

Whether an empirical, mechanistic, or spiked-sediment toxicity test approach is taken, the specific method will dictate the source data necessary. Despite the underlying basis of the approach, all data should be evaluated for relevance and reliability for deriving SQC. An ideal protocol will incorporate guidance on where and how to find data to ensure that the most up-to-date dataset is used for the calculation.

An extensive review of data sources and literature search protocols for deriving water quality criteria was conducted by TenBrook et al. (2009). As EqP approaches use WQC for derivation of SQGs, the relevant data sources are the same. In the United States, pesticide registrants must submit aquatic toxicity data to the USEPA and to the California Department of Pesticide Regulation (CDPR) if registering for use in California. Both agencies maintain databases of available pesticide toxicity data, which may be obtained on request (OPP Pesticide Ecotoxicity Database (2012) and CDPR Pesticide Data Index (CDPR 2012)). These databases comprise both aqueous and sediment toxicity test data. In addition, the USEPA maintains the ECOTOX database, which includes results from single-chemical toxicity studies for aquatic and terrestrial life published in peer-reviewed literature. The Danish WQC methodology recommends the LOGKOW (2000) database by Sangster Research Laboratories as a source of evaluated octanol-water partition coefficients, as well as literature searches using the BIOSIS database. Australia and New Zealand WQC derivation methodologies recommend collecting data from international criteria documents, the ECOTOX database, open literature, and review papers. Information on compound physicochemical properties can be obtained from the online hazardous substance data bank (HSDB; Verschueren 1983, most recent version Verschueren 2001 CD-ROM), and Hansch et al. (1995).

The Dutch protocol (RIVM 2001) for SQG derivation provides detailed information on relevant data sources for both ecotoxicity and physicochemical data. Recommended search topics include ecotoxicity data for all aquatic species (freshwater and saltwater), soil organisms, enzymatic activities, microbial processes, sediment-dwelling organisms (and birds and mammals if secondary poisoning is a concern), and partitioning coefficients. For biocide and plant protection products, environmental fate and toxicity data must be submitted to the Dutch Board for the Authorisation of Plant Protection Products and Biocides (Ctgb), and this information can be requested for review and guidance. The Dutch method states that only primary literature is to be used for SQG derivation. Kalf et al. (1999) proposed that ecotoxicity endpoints from the registration dossier and the scientific literature should be used together with environmental fate endpoints from the registration materials. Online bibliographic databases are useful sources of information: BIOSIS for ecotoxicity and physicochemical data, Chemical Abstracts for partitioning coefficient data, and TOXLINE for mammalian ecotoxicity data. Reliable sources for estimated or empirical physicochemical properties can be found in the handbooks of MacKay et al. (1999) and Boethling and MacKay (2000). Libraries recommended by the Dutch method include the Centre for Substance and Risk Assessment of the National Institute for Public Health and the Environment (RIVM 2012) and the Ctgb library (Ctgb 2012). The gray literature is searched only if time and budget allow, and the secondary literature should only be used to identify primary sources (Kalf et al. 1999). The Dutch method requires the literature to be reviewed as far back as 1970, or the first year of the database, so as to identify all available literature, particularly if no extensive review article identifying all primary sources has been published (RIVM 2001).

The USEPA database (OPP) includes acute and chronic aquatic toxicity data for both benthic and water column-dwelling species of varying sensitivities. The procedure and minimum data requirements for deriving the final chronic value are described by USEPA (1985). This methodology states that a "complete search, retrieval and review for any applicable data must be conducted, to locate all preexisting toxicity data." It is important to reexamine the FCV of a compound to ensure that the most up-to-date toxicity data are included. If no FCV is available for the chemical of concern, it is suggested that one be calculated, provided that the minimum data provisions are met as set forth by the USEPA (1985). Literature searches are also recommended as sources of toxicity data; however, as noted, the USEPA methodology does not specify a procedure to evaluate the quality of such literature.

The EU protocol is part of an overall risk assessment framework that calls for "the collection of all available information by manufacturers, importers and rapporteur" (ECB 2003). Little information is given as to where or how to find ecotoxicity data that is "complete and adequate" for deriving predicted no-effect concentrations for individual chemicals. The EU protocol states that test results from peer-reviewed journals are preferred, but quality review articles, summaries, and abstracts may be used as supporting materials (ECB 2003). There is no mention of specific sources of data.

The OECD and Ontario methods are based on equilibrium partitioning and do not include recommendations for data sources. The EqP method of the UK (Rowlatt et al. 2002) also lacks guidance on specific sources of data but requires that published literature, commercial databases, and unpublished data (e.g., manufacturer's data) be gathered for deriving environmental quality standards (Zabel and Cole 1999). The data is to be summarized and assessed for physicochemical properties, adequacy of methodologies, environmental fate and behavior, environmental concentrations, toxicology, and bioaccumulation (Zabel and Cole 1999).

The NOAA National Status and Trends Program has operated the national mussel watch and bioeffects assessment programs for US coastal waters since 1986. During this time, large amounts of sediment and bivalve tissue chemistry data have been collected for a suite of organic compounds and trace metals. The bioeffects assessment program has applied the sediment quality triad (sediment chemistry, sediment toxicity testing, and species diversity assessment) to identify and assess contaminant exposure effects across 40 regional studies. These monitoring efforts have yielded a large database of matching sediment chemistry and biological effects data (Long and Morgan 1990), which is referred to as a biological effects database for sediments (BEDS). Because of the ongoing nature of the program, methods for sample collection and analysis are standardized and well documented, and standard protocols have been developed for data quality control before entry into the database (Long and Morgan 1990). Agencies that derive SQGs by empirical approaches may lack adequate local data, and these jurisdictions often use the NOAA NSTP database for sediment assessments and BEDS development. The NOAA database is continuously updated as new information from these agencies becomes available. Australia and New Zealand (ANZECC and ARMCANZ 2000), Florida (MacDonald 1994), and Canada (CCME 1995) all follow the BEDS method. The state of Washington derived SQGs utilizing a local freshwater sediment quality database (FEDSQUAL) of matching sediment chemistry and biological effects data from both Oregon and Washington (Cubbage et al. 1997). The California empirical method (SWRCB 2011) does not include a detailed methodology for data and literature sourcing.

The Canadian method requires a comprehensive review and literature search for the compound of interest (CCME 1995). The physicochemical properties of the chemical are to be summarized, but the methodology does not include a description of the source of these data or guidance on how to judge data acceptable. Toxicological studies are to be found in the scientific literature and should be reviewed for quality according to the procedure outlined in the NSTP methodology. The literature review is to be used to summarize the chemical production and uses, known environmental fate data, sources into the aquatic environment, and to help evaluate and establish background concentrations. As the NOAA NSTP empirical approach does not incorporate bioavailability, the Canadian method recognizes the importance of including toxicological data from sediment exposures (CCME 1995).

To summarize, the approach taken will determine the data required to derive SQGs, but common for all approaches is the importance of achieving the most comprehensive high-quality dataset. This goal can only be achieved through detailed guidance on appropriate data sourcing, which should form an integral part of any methodology aimed to set chemical limits for any environmental compartment including sediment.

5.2 Physicochemical Data

Data on the physicochemical properties of a compound can be used as a starting point for understanding how a chemical will move and persist in the environment and which environmental compartments (viz., air, water, soil, sediment, and biota) are at risk of chemical exposure. Each SQG derivation methodology describes the types of physicochemical data to be gathered in different levels of detail.

5.2.1 Physicochemical Data Collection

In the Netherlands, the physicochemical data required include the International Union of Pure and Applied Chemistry (IUPAC) name, Chemical Abstracts Service (CAS) number, EINECS (European Inventory of Existing Commercial Substances) number, diagram of structural formula, empirical formula, molar mass, K_{ow}, water solubility, melting point, vapor pressure, Henry's law constant, and the acid dissociation constant (pK_a). In addition to the aforementioned parameters, K_d (referred to as K_p in RIVM 2001 or as $K_{s/l}$ in Kalf et al. 1999) and degradation rates (biotic and abiotic processes) should also be obtained.

The OECD (1995) calls for the collection of chemical structure, molecular weight, melting point, water solubility, K_{ow}, K_d, and pK_a. The bioconcentration factor (BCF), which represents uptake from the water column, may also be required if secondary poisoning is an issue. Experimentally determined BCFs are preferred over those estimated from the K_{ow} because BCF depends on other factors such as metabolism of the chemical within the organism.

The minimum physicochemical data requirement of the USEPA methodology is the K_{ow}, although an experimentally determined K_{oc} is preferred (Di Toro et al. 2002). The requirements of the UK (Rowlatt et al. 2002) and Ontario (Persaud et al. 1993) methodologies are similar. The EqP criterion calculation of these jurisdictions uses the K_{oc}, which may be estimated from the K_{ow}. The Ontario method requires at least three estimates of the K_d to set a SQG using the EqP approach (Persaud et al. 1993).

Empirical approaches mainly use matching sediment chemistry and biological effects data from field-collected sediments for calculating effects range concentrations. Compound physicochemical properties are thus not a requirement, and little guidance is provided for the collection of such data by the empirical methods (NOAA NSTP, Canada, Ontario, and California). It is important to note, however, that physicochemical data are crucial for understanding chemical transport and transformation processes in the sediment.

5.2.2 Quality Assurance for Physicochemical Data

It is extremely important to have accurate physicochemical data, particularly K_{ow} or K_{oc} values, because these are directly used for deriving SQGs in the EqP approach. Partition coefficients of highly insoluble chemicals, such as pyrethroids, are difficult to determine experimentally, and values for these compounds can vary by orders of magnitudes in the literature. To overcome this variability, the USEPA (Di Toro et al. 2002) and Dutch (RIVM 2001) EqP methods provide specific guidance to address data quality issues associated with partition coefficients required to calculate SQGs. The USEPA EqP method recommends using K_{ow}s from Karickhoff and Long (1995) and Long and Karickhoff (1996) to calculate the K_{oc} when available. When other literature-based K_{ow}s are used, these should be obtained from studies applying newer experimental methodologies such as the slow stir method (de Bruijn et al. 1989) and the generator column method (Woodburn et al. 1984). Site-specific K_{oc}s are not considered in the USEPA methodology. It is proposed, however, that sediment-water isotherms are generated and normalized for OC content during routine testing; such isotherms can be used for calculation of K_{oc} without additional testing (Di Toro et al. 2002). As EqP-derived SQGs are based on WQC, other physicochemical data quality issues would be identical to those used in WQC derivation and summarized by TenBrook et al. (2009).

The Dutch method specifies that chemical water solubility, Henry's law constant, log K_{ow}, and K_d should be gathered as background information for SQG derivation

(RIVM 2001). The method requires that the K_d be experimentally determined following the protocol for batch isotherm experiments for organics described by Bockting et al. (1993). All information related to the K_d is considered useful. The Freundlich exponent $(1/n)$ is particularly important as this corrects for the influence of increasing contaminant load on the adsorption isotherm (Mensink et al. 1995). Only K_ds with a Freundlich exponent between 0.7 and 1.1 should be used in the SQG calculation according to Kalf et al. (1999). The humus, organic matter, or organic carbon content must be reported along with the K_d. Temperature is an important variable to be considered when measuring equilibrium-based values such as K_d, water solubility, vapor pressure, and Henry's constant. The Dutch method considers the standard temperature for laboratory toxicity tests of 25°C to be appropriate. Additional information includes pH, cation exchange capacity, and the ability to calculate a mass balance from the data (preferably both water and sediment measured concentrations). If experimental K_{oc} data are lacking, values can be collected from the SRC database or handbooks (e.g., MacKay et al. 1999). If K_{oc} cannot be obtained from the literature, it can be estimated from K_{ow}. The recommended quantitative structure activity relationship (QSAR) for this calculation is the regression equation of Gerstl (1990):

$$\log K_{oc} = a \log K_{ow} + b, \qquad (2)$$

where a and b are constants for specific groups of chemicals as presented in the Dutch method (RIVM 2001).

During environmental breakdown, chemicals may form toxic metabolites, which can be relevant to set environmental quality criteria. The Dutch method applies chemical degradation information to assess whether it is appropriate to test the parent compound, the degradation product(s), or both for toxicity. As degradation via hydrolysis can be relatively fast, RIVM researchers developed a decision tree for testing of compounds with hydrolysis as the main breakdown route (Kalf et al. 1999; Mensink et al. 1995). Decisions are made based on the degradation half time (DT_{50}); that is, for DT_{50} values that are <4 h, tests are initiated with the metabolite(s), for DT_{50} equal to or exceeding 24 h, tests are initiated with the parent compound, and if DT_{50} values are between 4 and 24 h, expert judgment is used to determine whether to test the parent compound, the metabolite(s), or both (Kalf et al. 1999). This decision tree has been used for deriving harmonized maximum permissible concentrations in the Netherlands.

The EU method states that measured K_ds are preferred but may be estimated from K_{oc} or K_{ow} (ECB 2003). Solid-water partition coefficients may be obtained from direct measurement, simulation testing, measured by adsorption studies or the high-pressure liquid chromatography (HPLC) method, or estimated from the K_{ow} using QSARs. The OECD (1995) requires the K_{ow} to be determined using the slow stir or generator column methods for compounds with log K_{ow} > 5. Expert evaluation of the values is also recommended by the OECD to ensure high data quality. Ontario's EqP method states that both measured and calculated K_{ow}s may be used to determine K_{oc} but that experimentally derived K_{oc} data should be used for SQG calculation whenever possible (Persaud et al. 1993). At least three estimates of partition coefficients are required to set a SQG using the EqP approach. If less than three values are available, the SQG is considered tentative.

If bioconcentration or bioaccumulation is of concern (i.e., log $K_{ow} > 3$), the BCF and/or the sediment (or soil) accumulation factors can be obtained or calculated from experimental studies. Dutch methods recommend that the selected studies include information such as species, species properties (e.g., age, size, weight, lifestage, sex, if known), test type (semi-static, static, continuous flow, intermittent flow), water properties (hardness or salinity), exposure time and concentration, time to equilibrium, and dry to wet wt. ratio (RIVM 2001). The USEPA provides criteria, to which studies must adhere, to use associated BCFs for deriving final tissue residue values (USEPA 1985, detailed in TenBrook et al. 2009). Briefly, BCFs must be based on concentrations measured in tissue and test solution in flow-through experiments conducted at steady-state conditions. The percent lipid in tissue must be reported for lipophilic compounds, and chemical concentrations should be reported on a wet wt. basis. Where more than one BCF are available for similar exposure conditions, the geometric mean of all values across species is to be applied.

5.3 Ecotoxicity Data

Available ecotoxicological information generally spans short-term to long-term chemical exposure studies that employ a variety of endpoints (lethal, sublethal, biochemical). Ecotoxicity data are generated through both single- and multiple-species tests performed in both laboratory and field studies. TenBrook et al. (2009) details the different types of ecotoxicity data found in the literature. Data include values representing various degrees of toxicity including lethal and effects concentrations (LC_x/EC_x) derived through a range of aqueous and sediment exposure toxicity tests. While internationally accepted protocols for aquatic toxicity testing have been in place for over a decade, many of the standard methods for sediment toxicity testing are still undergoing development and/or validation as new research on appropriate test organisms, endpoints, and variation across sediments is finalized. Many of the test methods developed by the USEPA, OECD, and ASTM are therefore considered guidelines that are yet to be fully validated. Jurisdictions using the EqP approach often rationalize the implementation of such an indirect approach by this lack of standard protocols for sediment toxicity testing and the consequent data shortage.

5.3.1 Acute Versus Chronic Exposure

Sediment quality criteria aim to be protective of aquatic life during both short-term transient exposures and long-term continuous exposures. Different types of toxicity tests have been developed to assess both long- and short-term effects of sediment contaminants on benthic organisms. Acute tests are conducted through short-term exposure and generally measure mortality or immobility, while chronic tests are reflective of long-term exposure and endpoints such as survival, growth, emergence, and reproduction. The USEPA Office of Research and Development (USEPA 2000a), USEPA Office of Prevention, Pesticides, and Toxic Substances (USEPA 1996a, b, c, d, e, f, g, h, i, j), OECD (1992, 2004a, b, c, 2007, 2008), Environment Canada (1997a, b), and ASTM (2004, 2006a, b, 2007a, b, 2008a, b, c, d, 2010) have all developed standardized methods for sediment toxicity testing.

The current ASTM method for toxicity testing of sediment-associated contaminants with freshwater invertebrates specifies that short-term (acute) tests are of 10-day duration and include both survival and growth endpoints (ASTM E 1706-05, ASTM 2008a). Long-term (chronic) tests for *Hyalella azteca* should be conducted over 42 days with endpoints measured at 28, 35, and 42 days including survival, growth, and reproduction, while a long-term test for *Chironomus dilutus* (formerly *C. tentans*) entails a 20-day life-cycle test with endpoints of growth, survival, reproduction, and emergence. The USEPA Office of Prevention, Pesticides and Toxic Substances protocol OPPTS 850.1735 also describes a method for acute sediment toxicity testing with the freshwater organisms *H. azteca* and *Chironomus tentans* (USEPA 1996c).

Definitions of acute and chronic test durations are not always consistent across SQG methodologies. Test durations should reflect species life-cycle durations, which can vary substantially among taxa. A USEPA report defined, however, acute laboratory sediment toxicity tests lasting 10–14 days as being acute and tests of 21–60 days duration as being chronic (Ingersoll and MacDonald 2002). These definitions are consistent with the ASTM sediment toxicity test guidelines for invertebrates described above. The definitions of acute and chronic test durations given in the Dutch method differ somewhat from those stated by the USEPA. The Dutch guidelines give the following taxa-specific definitions for exposure: For algae and protozoa, tests of 3–4 days are considered chronic (longer exposure is accepted if growth is still in the exponential phase); for Crustacea and Insecta, test durations of 48 or 96 h are considered acute; for Pisces, Mollusca, and Amphibia, 96-h tests are considered acute, while 28-day early lifestage tests are considered to be chronic (RIVM 2001).

The Canadian methodology does not specifically define acute and chronic toxicity but does state that "ideally, SQGs should be developed from detailed dose–response data that describe the acute and chronic toxicity of individual chemicals in sediment to sensitive lifestages of sensitive species of aquatic organisms" (CCME 1995). In this method, the minimum requirements of four data must include at least two chronic tests covering partial or full life cycles.

Washington State defines an acute test as a "measurement of biological effects using surface sediment bioassays that are short in duration compared to the life-cycle of the test organism" (WAC 1995). Acute effects include mortality, larval abnormalities, or other endpoints deemed appropriate. Chronic tests are defined as measurements of biological effects using surface sediment bioassays over a period not less than one complete life cycle of the test organism. The term chronic also includes evaluations of indigenous field organisms for long-term effects, as well as the effects of biomagnification and bioaccumulation. "Chronic effects may include mortality, reduced growth, impaired reproduction, histopathological abnormalities, adverse effects to birds and mammals or other endpoints determined appropriate" (WAC 1995).

Unlike exposures in the water column, benthic invertebrates are typically exposed to sediment contaminants for extended periods of time because of the accumulative nature of bedded sediments (Ingersoll and MacDonald 2002). In a study by the USEPA (2000b), long-term (chronic) sediment toxicity tests with growth and survival endpoints tended to be more sensitive than short-term (acute) tests. Based on these findings, Ingersoll and MacDonald (2002) recommended that chronic toxicity tests are more relevant for predicting effects in aquatic ecosystems and should be used to assess effects of contaminated sediment on aquatic organisms.

5.3.2 Hypothesis Tests Versus Regression Analysis

Two main options exist for analyzing ecotoxicological data; these are regression analysis and hypothesis testing. In regression analysis, concentration-effect data are plotted to derive a regression equation, from which it is possible to predict the concentration-effect pair for any given effect level (i.e., LC/EC_x) (Stephan and Rogers 1985). Hypothesis testing compares effects data for treatment and control groups to elucidate the concentration level that causes a statistically significant difference from the control (Stephan and Rogers 1985). The highest concentration for which responses are not statistically significantly different from the control is termed the no observed effect concentration or level (NOEC or NOEL),whereas the lowest concentration causing a statistically significantly different response compared to the control is referred to as the lowest observed effect concentration or level (LOEC or LOEL). The maximum acceptable toxicant concentration (MATC) is the geometric mean of the NOEC and LOEC. Hypothesis testing is typically used for chronic tests covering full, partial, or early lifestages, while regression analysis is more commonly used with acute exposure tests. The advantages and disadvantages of the two statistical approaches are discussed in relation to WQC by TenBrook et al. (2009), and similar concepts apply for SQC. Regression methods are the overall preferred tool, but chronic threshold values derived by hypothesis testing are acceptable to make use of valuable limited chronic exposure data.

5.3.3 Single-Species (Laboratory) Versus Multispecies (Field/Semi-field) Data

Single-species laboratory tests are standardized and relatively easy to interpret. Data are used directly in the SSTT approach for criteria derivation, indirectly in the EqP approach for verification, and as supplementary data in co-occurrence datasets for empirical approaches. Although multispecies laboratory tests, field studies, mesocosm, and microcosm tests better approximate natural ecosystems, these types of tests are less standardized, often lack replication, and are difficult to interpret

(TenBrook et al. 2009). As also concluded for WQC, multispecies studies are useful supplementary tools, but unlikely to be used as sole data sources for sediment quality criteria due to comparatively limited cost-effectiveness, reproducibility, and reliability (TenBrook et al. 2009).

Several methodologies propose the use of multispecies field studies in the final stages of criteria derivation. Both the Netherlands (RIVM 2001) and the EU (ECB 2003) methods recommend comparing the results of multispecies field studies to criteria derived from single-species laboratory toxicity tests. These methods note that it is more difficult to interpret field studies than controlled laboratory tests, due to the variation in test parameters and exposures for field studies. The Dutch method offers a set of criteria by which field data can be evaluated. The Dutch method further advises that NOECs from multispecies tests should be compared to the derived criteria to establish if these are at risk of being underprotective of ecosystems (RIVM 2001).

5.3.4   Traditional Versus Nontraditional Endpoints

Traditional endpoints of standard test methods include responses such as survival, growth, and reproduction that are clearly linked to population-level effects. Traditional endpoints are recommended for use in all reviewed methodologies, although nontraditional endpoints are used by some methods on a case-by-case basis. Examples of nontraditional endpoints include endocrine disruption, enzyme induction and inhibition, behavioral and histological effects, stress protein induction, altered RNA and DNA levels, mutagenicity, and carcinogenicity (TenBrook et al. 2009).

The Canadian guidelines advocate that toxicity tests should follow standard methods for assessing ecologically relevant endpoints, which include survival, growth, reproduction, and developmental effects (CCME 1995). Similarly, the Netherlands methodology only employs data for population-level endpoints, such as survival, growth, and reproduction (RIVM 2001). Examples of reproductive effects included in the Dutch protocol are histopathological effects on reproductive organs, spermatogenesis, fertility, pregnancy rate, number of eggs produced, egg fertility, and hatchability (RIVM 2001). The Dutch method further recommends that data for additional endpoints may be compared to the derived criteria (RIVM 2001). This additional information can help ensure that the derived criteria are protective, particularly for chemicals that have specific modes of action (e.g., phthalates, which are suspected endocrine disruptors).

The EU method endorses using studies in which standard endpoints are not applied (i.e., related to survival, growth, or reproduction) if an expert judges that such data can be included with standard endpoints (ECB 2003). Standard endpoints in the EU protocol include emergence, sediment avoidance, and burrowing activity, in addition to survival, growth, and reproduction (ECB 2003). Nontraditional endpoints used in the EU may include other behavioral effects, photosynthesis, or cellular and subcellular effects.

5.3.5 Data Estimated from Interspecies Relationships

For many species likely to be present in ecosystems, no toxicity data are available. To increase the number of species represented in SQC without performing additional toxicity tests, it has been proposed that toxicity values can be estimated based on interspecies relationships. For aquatic toxicity data, the USEPA has developed a program to estimate acute toxicity for untested species based on available data for more common test species (Raimondo et al. 2010). Unfortunately, the USEPA program does not include data for sediment exposures, and thus, this software is not applicable for derivation of sediment quality criteria. No alternative interspecies correlation approaches were identified that included sediment exposures.

5.3.6 Ecotoxicity Data Quality Assurance

Data quality is generally assured by means of standardized protocols for toxicity testing. In the development of SQGs, different types of protocols have been used, depending on the experimental question being addressed. This section reviews sediment toxicity tests used for empirical approaches, as well as spiked-sediment toxicity tests applicable for deriving SQC using species sensitivity distributions or assessment factors. Strategies to evaluate if individual studies comply with standard methods are also reviewed.

Standard Methods

A list of selected current standard sediment toxicity testing methods and related protocols is provided in Table 4. The Dutch method highlighted a lack of internationally accepted protocols for toxicity testing in sediment-water systems (RIVM 2001). Since this publication, standard methods have become available from several jurisdictions, and these are further described below.

The ASTM E 1706-05 method entitled "Standard test methods for measuring the toxicity of sediment-associated contaminants with freshwater invertebrates" describes 10-day testing protocols for *Hyalella azteca* and *Chironomus dilutus*, using whole sediments from field-collected or laboratory-spiked sediments (ASTM 2008a). This method also provides guidance on conducting short-term sediment toxicity tests with *Chironomus riparius*, *Daphnia magna*, *Ceriodaphnia dubia*, *Hexagenia* spp., *Tubifex tubifex*, and *Diporeia* spp. Further, ASTM 1706-05 provides instructions for long-term sediment toxicity testing, initially developed for testing with *H. azteca* and *C. dilutus* but applicable to all of the above-named organisms. Bioaccumulation tests with sediment-associated contaminants are addressed in a separate method (ASTM 1688-10), which details a 28-day study with the oligochaete *Lumbriculus variegatus* (ASTM 2010). According to ASTM 1706-05, future method updates will include results of research into the use of formulated sediment, refinement of sediment-spiking procedures, and evaluation of endpoint sensitivities (ASTM 2008a).

Table 4 List of selected sediment toxicity testing methods and related protocols

Method Source	Method ID	Title
USEPA 1996a	850. 1735 (S)	Whole sediment acute (or chronic) toxicity: invertebrates, freshwater
USEPA 1996b	850. 1740	Whole sediment acute toxicity: invertebrates, marine
USEPA 1996c	850. 1790	Chironomid sediment toxicity test
USEPA 1996d	850. 1800	Tadpole/sediment subchronic toxicity test
USEPA 1996e	850. 1850	Aquatic food chain transfer
USEPA 1996f	850. 1900	Generic freshwater microcosm test, laboratory
USEPA 1996g	850. 1925	Site-specific aquatic microcosm test, laboratory
USEPA 1996h	850. 1950	Field testing for aquatic organisms
USEPA 1996i	850. 1010	Aquatic invertebrate acute toxicity test, freshwater, daphnids
USEPA 1996j	850. 1075	Fish acute toxicity test, freshwater, and marine
ASTM (2008a)	E 1706-05 (2008)	Standard tests method for measuring the toxicity of sediment-associated contaminants with freshwater invertebrates
ASTM (2008b)	E 1367-03 (2008)	Standard test method for measuring the toxicity of sediment-associated contaminants with estuarine and marine invertebrates
ASTM (2008c)	E 1391-03 (2008)	Standard guide for collection, storage, characterization, and manipulation of sediments for toxicological testing and for selection of samplers used to collect benthic invertebrates
ASTM (2008d)	E 1525-02 (2008)	Standard guide for designing biological tests with sediments
ASTM (2010)	E 1688-10	Standard guide for determination of the bioaccumulation of sediment-associated contaminants by benthic invertebrates
ASTM (2006a)	E 2455-06	Standard guide for conducting laboratory toxicity tests with freshwater mussels
ASTM (2007b)	E 2591-07	Standard guide for conducting whole sediment toxicity tests with amphibians
ASTM (2006b)	E 1295-01 (2006)	Standard guide for conducting three-brood, renewal toxicity tests with *Ceriodaphnia dubia*
ASTM (2004)	E 1193-97 (2004)	Standard guide for conducting *Daphnia magna* life-cycle toxicity tests
USEPA (1994b)	EPA 600-R24-024	Methods for measuring the toxicity and bioaccumulation of sediment-associated contaminants with freshwater invertebrates
USEPA (1994a)	EPA 905-R94-002	Assessment guidance document, Great Lakes Program (EPA 600-R94-025; EPA 600-R99-064)
OECD (2004a)	218	OECD No. 218: Sediment-water chironomid toxicity using spiked sediment
OECD (2004b)	219	OECD No. 219: Sediment-water chironomid toxicity using spiked water
OECD (2007)	225	OECD No. 225: Sediment-water *Lumbriculus* toxicity test using spiked sediment
OECD (1992)	210	OECD No. 210: Fish, early-life stage toxicity test
OECD (2004c)	202	OECD No. 202: *Daphnia* sp. acute immobilization test
OECD (2008)	211	OECD No. 211: *Daphnia magna* reproduction test

ASTM = American Society for Testing and Materials, OECD = Organisation for Economic Co-operation and Development, OPPTS = Office of Prevention, Pesticides and Toxic Substances, USEPA = United States Environmental Protection Agency

The ASTM 1706-05 method includes freshwater organisms with different feeding and habitat requirements (Table 5). As data for *Hexagenia* spp., *T. tubifex*, and *Diporeia* spp. are comparatively less robust (relative to data available for *H. azteca* and *C. dilutus*), the protocols for these species are currently considered to be guidelines only (ASTM 2008a). In addition to this method for invertebrates, an ASTM protocol is available for conducting sediment toxicity tests with amphibians (ASTM E 2591-07, ASTM 2007b). The amphibian protocol is, however, also considered to be only a guideline, rather than an official test method. The lack of standard protocols for sediment toxicity testing toward a wider range of benthic community members is a major limitation of SSTT derivation methodologies, but it should be noted that some guidance is available and further standard methods development appears to be underway.

The USEPA Office of Prevention, Pesticides and Toxic Substances (OPPTS), which was recently renamed the Office of Chemical Safety and Pollution Prevention (OCSPP), has developed harmonized test guidelines for registering pesticides and toxic substances. The goal of harmonization is to minimize the variation among the testing procedures used to fulfill data requirements for the Toxic Substance Control Act and the Federal Insecticide, Fungicide, and Rodenticide Act. These harmonized procedures combine testing guidance from the OECD, USEPA Office of Pollution Prevention and Toxics (OPPT), and the USEPA Office of Pesticides Programs (OPP). The OPPTS 850.1735 method (USEPA 1996c, Sect. 5.3.1) was adapted from the USEPA protocol entitled "Methods for measuring the toxicity and bioaccumulation of sediment-associated contaminants with freshwater invertebrates" (EPA 600-R24-024, USEPA 1994b). This method represents the harmonized version of the USEPA's ecological effects test methods for sediment tests with freshwater invertebrates. Although the OPPTS series 850 guidelines (including USEPA 1996a, b, c, d, e, f, g, h, i, j) are not considered final, these may be applied for protocol development.

Fleming et al. (1996) performed intra- and inter-laboratory comparisons of sediment toxicity tests and found that most of the variability between test results could be attributed to sediment-spiking procedures. It was proposed that standardized spiking methods would need to address sediment heterogeneity, appropriate characterization of the variables controlling sorption and bioavailability, equilibration, and aging times. Fuchsman and Barber (2000) suggest a simpler solution to address sediment-spiking issues, which involves pre- and posttest measurement of sediment concentrations to confirm test exposures and assess chemical loss over the exposure duration. Sediment-spiking instructions recommended by the ASTM include confirmation of sediment concentrations before toxicity test initiation (ASTM E 1367-03, ASTM 2008b).

Data Relevance and Reliability

A detailed description of the processes by which aquatic ecotoxicity data are judged for quality in the Netherlands, UK, Canada, and Australia/New Zealand is presented

Table 5 Species included in the ASTM 1706-05 method, representing freshwater organisms with different feeding and habitat requirements

Species	Phylum (subphylum or class)/family	Habitat	Feeding characteristics	Other sediment-related characteristics
Hyalella azteca	Arthropoda (Crustacea)/Hyalellidae (amphipod)	E	Partial subsurface deposit feeder	Wide tolerance of sediment grain size
Daphnia magna	Arthropoda (Crustacea)/Daphniidae (water flea)	WC	Filter feeder	
Chironomus riparius	Arthropoda (Insecta)/Chironomidae (midge)	E	Filter feeder/surface deposit feeder	Larvae burrow into sediment (direct contact); wide tolerance of sediment grain size
C. dilutus (formerly *C. tentans*)	Arthropoda (Insecta)/Chironomidae (midge)	E	Filter feeder/surface deposit feeder	As above
Ceriodaphnia dubia	Arthropoda (Crustacea)/Daphniidae (water flea)	WC	Filter feeder	
Tubifex tubifex	Annelida/Tubificidae (Oligochaete)	E, I	Subsurface deposit feeder	Tolerant of variation in sediment particle size and organic matter proportion; key species in aquatic food chain and active in bioturbation
Hexagenia spp.	Arthropoda (Insecta)/Ephemeridae (mayfly)	E, I	Surface particle collector	Nymphs burrow into sediment (direct contact); prefers fine sediments rich in organic matter
Diporeia spp.	Arthropoda (Crustacea)/Pontoporeiidae (amphipod)	E, I	Deposit feeder	Relatively insensitive to grain size
Lumbriculus variegatus	Annelida/Lumbriculidae (oligochaete)	E, I	Subsurface deposit feeder	Inhabits a wide variety of sediment types

Habitats: *I* infaunal, *E* epibenthic, *WC* water column

by TenBrook et al. (2009). EU protocols only accept ecotoxicological data for deriving criteria if these are adequate, complete, and have been acquired using standardized, internationally accepted protocols (Table 4). Test procedures and designs that deviate from the standard are usually reviewed against measures that use best professional judgment (ECB 2003). The adequacy of a study is determined by its reliability and relevance. Reliability is evaluated by the quality and description of the test method used (i.e., ASTM, OECD, and/or USEPA test methods with good laboratory practice). A relevant study is designed to test appropriate endpoints under relevant conditions and a test compound that is representative of the chemical being assessed.

In the Netherlands, the quality of ecotoxicological data is ranked according to a reliability index (RIVM 2001). A score of 1 applies to a study using a methodology that is in accordance with accepted international test guidelines and/or Mensink et al. (1995), a score of 2 indicates less accord with accepted test methods, and a score of 3 applies to data that is of inadequate reliability to calculate the maximum permissible concentration. The Dutch method further requires that the purity of the test substance is at least 80%. Data generated from a less pure substance may not be included directly for SQG derivation, but can be used as supporting information. An exception is made for granulates and wettable powders of purity between 20 and 80% if absences of carrier toxicity have been established. Studies using polluted animals are rejected, and aquatic studies must demonstrate at least 80% recovery of the test substance, which may not be tested at concentrations exceeding ten times its aqueous solubility and/or with a solvent concentration exceeding 1 mL/L. Sediment toxicity test requirements are less detailed than those for aquatic studies. However, the Dutch methodology recommends evaluation of certain test parameters, including sediment characteristics (% organic carbon, particle size distribution, field, or standard sediment), the amount of sediment and water used, test method (static or flow-through), spiking method, measured chemical concentrations (i.e., in sediment and/or water) if at steady state, system description (i.e., suspended or bedded sediment), and exposure route.

In Canada, "accurate and precise" data generated using standard sampling techniques and appropriate test methods are important to maintain data integrity. Sediment characteristics also need to be determined (e.g., grain size, total organic carbon (TOC)) for interpretation of biological effects. The CCME (1995) specifically states that test methods should include light and dark cycles and should verify the condition of test organisms throughout the duration of exposure. Chemical concentrations should be measured in water at the beginning and end of the test, in both overlying water and sediment compartments. Data on the health and survival of the test organism before exposure should be documented for at least 1 week prior the start of the test. Test organisms should not be used if significant mortality has occurred during this time frame.

The NSTP empirical approach used in Canada also requires quality ecotoxicity data for incorporation into a BEDS. A detailed description of the evaluation of ecotoxicity data is provided by CCME (1995). To ensure high-quality data, sampling, storage, and handling of sediments should be consistent with standard protocols (e.g., ASTM 2008a, b, c, d; Environment Canada 1994; Loring and Rantala 1992). In terms of holding time and storage, sediments should be tested within 2 weeks of collection and must not be frozen. Accepted toxicity tests are those that follow standard protocols (e.g., ASTM 1990a, b; Environment Canada 1992a, b, c, 1995). Nonstandard toxicity testing methods should be evaluated on a case-by-case basis. Sediment chemical concentrations must be measured by appropriate analytical techniques at a number of time points dependent on the chemical and test duration. Nominal concentrations are not acceptable (CCME 1995). The sediment should be characterized for TOC, particle size distribution, acid volatile sulfides, pH, redox conditions, and sediment type, while overlying water should be characterized for pH, dissolved oxygen, total suspended solids, suspended and dissolved organic carbon, and water hardness (and/or alkalinity) or salinity. Embryonic development, early lifestage survival, growth, reproduction, and adult survival are preferred endpoints, although other endpoints related to organism pathology or behavior (avoidance, burrowing) may also be considered. It is obligatory to monitor control survival and response, which must be within acceptable limits and appropriate for the lifestage of the organism tested. Static, static renewal, or flow-through aquatic ecotoxicity tests may be included in the dataset for sediment assessment using the Canadian NSTP methodology. Maintenance of adequate environmental conditions must be demonstrated for the test duration. Data that lack sufficient information to assess the adequacy of the test design, procedures, and/or results must be excluded from SQG derivation.

The California State Water Resources Control Board (SWRCB) requires that all test methods adhere to USEPA or ASTM methodologies or obtain approval by the State and Regional Water Boards (SWRCB 2011). In Washington, a quality assurance grade has been applied to each piece of data entered into the database and made available for reference. The grade (A–F) is assigned to each investigation based on protocols conducted and presented in the final report. The grade may not be directly representative of data quality but reflects the quality of the amount and types of quality assurance procedures completed in the investigation (Cubbage et al. 1997).

5.3.7   Required Quantity of Ecotoxicity Data

A full review of the data quantity required for calculation of WQC used in the EqP approach is presented by TenBrook et al. (2009). Direct SQG derivation methods that utilize data from sediment toxicity tests are, albeit limited, the focus of the current review. The Canadian SSTT approach outlines that the minimum dataset for deriving freshwater SQGs must include at least four studies on at least two or more sediment-residing invertebrate species found in North American waters (CCME 1995). One benthic arthropod and benthic crustacean species must be included, and at least two of the studies must be partial or full life-cycle tests that consider ecologically relevant endpoints (e.g., growth, reproduction, developmental effects). It is also recommended that "ecologically relevant species" are the focus of the data review (CCME 1995). Alternatively, the Canadian NSTP-based approach for interim sediment quality guideline derivation requires at least 20 entries into the effects and no-effect dataset (CCME 1995). The Dutch method states that when sediment toxicity data are lacking, aquatic toxicity data can be used to indirectly calculate the sediment maximum permissible concentration via the EqP approach (RIVM 2001). Similar to the Canadian NSTP approach, Ontario's use of the screening level concentration (SLC) empirical approach provides guidance on effects and no-effect database construction (Persaud et al. 1993). The range of concentrations entered into the database should span two orders of magnitude and comprise both heavily contaminated and relatively clean sites. At least 75% of the database entries must represent benthic infaunal species, with proper taxonomic identification to at least the genus level. A minimum of ten observations are required to set a species SLC (SSLC), and at least 20 different SSLCs are required for SLC calculation (Persaud et al. 1993).

The SWRCB method for California requires completion of a minimum of one short-term survival test and one sublethal test for each sediment sample collected from each station (SWRCB 2011). Short-term tests should comprise 10-day whole sediment exposures testing survival in acceptable test organisms tolerant of the sample salinity and grain size characteristics (i.e., *Eohaustorius estuarius*, *Hyalella azteca*, *Leptocheirus plumulosus*, *Rhepoxynius abronius*). Acceptable sublethal testing methods are whole sediment 28-day exposures, in which growth in *Neanthes arenaceodentata* is measured. A 48-h sediment-water interface exposure test using embryo development as the endpoint in *Mytilus galloprovincialis* is also accepted for sublethal testing. Sediment toxicity results are compared and categorized as nontoxic, low toxicity, moderate toxicity, and high toxicity relative to control performance (SWRCB 2011). The average of all test responses determines the final line of evidence category, and if the average falls between two categories, the higher response category is to be selected (SWRCB 2011).

In a study concerning the Great Lakes region, Burton et al. (1996) proposed that for field bioassays to adequately detect sediment toxicity, the test design should comprise 2–3 assays of various combinations of *Hyalella azteca*, *Chironomus tentans*, *Chironomus riparius*, *Ceriodaphnia dubia*, *Daphnia magna*, *Pimephales promelas*, *Hexagenia bilineata*, *Diporeia* spp., *Hydrilla verticillata*, and *Lemna minor*.

5.4   Quantitative Structure Activity Relationships (QSARs)

A QSAR is a mathematical model that describes the relationship between a compound's structure and its toxicity. TenBrook et al. discussed the role of QSARs in filling data gaps for WQC derivation (TenBrook et al. 2009). With respect to sediment quality criteria, QSARs may facilitate prediction of K_{oc} from K_{ow} (RIVM 2001). For pesticides with specific modes of action, such as the pyrethroids and organophosphates, QSAR models are still in the early stages of development but may be of use in the future (Zvinavashe et al. 2009).

5.5 Data Combination and Exclusion

Final data processing may entail combining and/or excluding data when multiple data exist for a single species. While many criteria derivation methodologies have defined guidelines for processing of aquatic data, only the Dutch method specifically contains such guidelines for sediment data. Most other methods simply assume that multiple data issues will not be encountered for benthic species, either because multiple data will not exist or because aquatic data are used to derive SQC via the EqP approach. Most WQC guidelines recommend using the geometric mean over the arithmetic mean as the most appropriate estimate across multiple data (TenBrook et al. 2009).

In the Netherlands, for a given chemical and toxicity value (e.g., LC_{50} vs. NOEC), toxicity data are selected to achieve one single reliable value for each species (RIVM 2001). To exclude or combine data, the following guidance is given: (1) The geometric mean of multiple data based on the same endpoint should be calculated for each species; (2) if data exist for multiple endpoints for a given species, the most sensitive endpoint is selected; and (3) if data exist for different lifestages of a given species, the most sensitive lifestage is selected. The Dutch guidelines also describe a method for converting chronic data to NOECs (detailed in TenBrook et al. 2009).

The EU method defines how to reduce data for the aquatic compartment; however, it assumes that data refinement is unnecessary for benthic organisms as multiple data are not expected to be available for any one species (ECB 2003). If a sediment criterion is calculated by the EqP approach using aqueous data, data reduction procedures would follow those for aquatic guidelines (outlined in TenBrook et al. 2009). These EU data refinement approaches may also be useful for benthic data, in cases where regulators are presented with multiple data for one species. As the OECD (1995) employs the EqP approach for SQC derivation, this method does not offer guidance for reduction of sediment data.

SSTT data are likely to be few. Yet, for the most common test species, such as *Hyalella azteca* and *Chironomus dilutus*, multiple data may exist and future methods should disclose guidance on how to select the most appropriate endpoint or duration for use in criteria derivation. Similar guidance as defined in WQC methods may be useful for refining multiple data for a given endpoint/species combination for SQC.

6 Criteria Calculation

The goal of this section is to present the various existing approaches for deriving sediment quality criteria. The review focuses on EqP and SSTT methodologies, including the assessment factor (AF) and species sensitivity distribution approaches. Although there are limitations to the empirical approaches for deriving single numeric criteria (Sect. 2.2), a discussion of these practices is also included. Issues related to mixture toxicity, bioaccumulation, secondary poisoning, threatened and endangered species, harmonization of criteria across environmental compartments, and data utilization are relevant for criteria derivation, and the importance of these factors are discussed throughout this section.

6.1 Exposure Considerations

To establish suitable sediment quality criteria, it is important to consider the exposure factors that affect sediment toxicity. Relevant factors of exposure include the magnitude, duration, and frequency of exposure as well as exposure routes and sediment/particulate characteristics, which contribute to bioavailability. In this section, we summarize the considerations made for the different methodologies, in regards to sediment-associated exposure, as well as the most recent research on the topic.

6.1.1 Magnitude, Duration, and Frequency

Exposures to sediment contaminants vary in magnitude, duration, and frequency, depending on the particular environmental conditions. TenBrook et al. (2009, 2010) proposed to define water quality criteria in terms of magnitude, duration, and frequency to determine exceedances, which closely follows the guidance in the USEPA (1985) methodology; the discussion below will address these aspects with regard to sediment quality criteria. TenBrook et al. (2009) give the example that "a criterion designed to protect against ongoing, chronic toxicant exposure that is stated in terms of magnitude only will be overprotective in cases of brief, mild excursions above the criterion, but will be underprotective in cases of brief, large excursions." To arrive at appropriate criteria, two approaches can be taken with respect to magnitude, duration, and frequency: (1) "Incorporate some combination of magnitude, duration and frequency in each criterion statement" or (2) "derive the magnitude only and leave duration and frequency determinations to site-specific management decisions" (TenBrook et al. 2009). All the reviewed methodologies address the numeric magnitude of a SQC; however, the duration and frequency components are only considered indirectly through guidance on compliance monitoring.

Exposure duration is an important consideration for areas that experience regular short-term toxic pulses, such as the Sacramento and San Joaquin River basins (Bailey et al. 2000; Dileanis et al. 2003; Kratzer et al. 2002), which regularly receive runoff from rain events and agricultural discharges. Chronic effects predictions are complicated for pulse exposures because long-term effects are determined by details of the specific exposure scenario in question (Forbes and Cold 2005). Although brief exposures to sediment-bound pesticides that cause short-term adverse effects to benthic organisms have been reported in some studies (Balthis et al. 2010; Hose et al. 2002), it is not clear if short-term exposures to deposited sediment contaminants will cause long-term adverse effects (Forbes and Cold 2005). A further consideration, however, is that repeated pulses may not allow full recovery after exposure

(Wallace et al. 1989). Together, the magnitude and duration aspects address the differences between pulse emissions and ongoing chronic exposure that result from the accumulation of residues in sediments. In WQC, the aspect of duration has often been addressed by deriving two separate criteria of different magnitude (i.e., acute and chronic criteria), while all SQG methods recommend deriving a single (typically chronic) value. The USEPA EqP methodology, for example, determines a single chronic SQG, utilizing only the final chronic value of the WQC, although acute values are available. A similar approach is taken by EqP methodologies of other jurisdictions (the Netherlands, EU, Ontario, France, OECD, and UK). One explanation for prioritizing chronic exposure data is that chronic values and hence, the resulting magnitude of criteria should be reflective of long-term exposures and in that sense, indirectly incorporate duration. Furthermore, the USEPA (2003a, 2003b) states that the duration and frequency components are irrelevant for SQGs, because it is expected that the concentration of sediment contaminants will be relatively stable over time, leading to chronic, relatively constant exposures of benthic species.

The frequency of exceedance component is designed to ensure that an ecosystem impacted by an excursion of the criterion has time to fully recover before another excursion may occur, because it is assumed that adverse effects may be compounded by multiple excursions without recovery. Toxicological studies of pesticide-contaminated sediment indicate that recovery times are dependent on many variables and can vary from as little as 1 month up to 3 years (Balthis et al. 2010; Caquet et al. 2007; Hatakeyama and Yokoyama 1997; Woin 1998; Yasuno et al. 1982). Frequency of exceedances may be the most difficult aspect to address for sediment contaminants as these can accumulate over time. Sampling in the same location over time may not reveal new inputs, but instead may confound results with residual pesticides, for which exceedances were previously recorded. All of the reviewed methodologies give instructions on deriving the magnitude of criteria but leave the duration and frequency components to site-specific judgment by environmental managers.

Sediment quality guidelines are generally used to complement existing sediment assessment tools, assess sediment contamination, and serve as targets for maximum contaminant loading in a water body (Di Toro et al. 2002). SQGs are not currently used in regulatory contexts alone, and as a result, there is little to no discussion regarding the allowable duration and frequency of SQG exceedance in the current methodologies. In most SQG methods, exceedances of magnitude are used as triggers for further study of the contaminant and for development of management practices designed to reduce chemical loadings. To account for the duration and/or frequency of sediment contaminant exposures, environmental managers may incorporate these aspects in the design of monitoring programs. The Dutch and German methods, for example, use the 90th percentile of annual compliance monitoring data based on the concentration in suspended particles (Germany and the Netherlands) or bedded sediments (the Netherlands only), for which the duration and frequency of exceedances depend on the sampling design (Lepper 2002). The EU is debating whether to base compliance of guidelines on the arithmetic mean or on the 90th percentile of the annual levels monitored in suspended particulate matter (Lepper 2002). Using the suspended particulate matter in compliance monitoring avoids the issue of sampling accumulated contaminants versus newly deposited contaminants, although resuspension of bedded sediments could still confound monitoring if sampling sites have high resuspension fluxes.

In summary, the magnitude of SQGs is clearly addressed by existing methodologies, while the duration and frequency components are only indirectly accounted for. The duration component may be incorporated using chronic data for deriving SQGs that are likely to be protective of long-term exposures, which are more representative of contaminated sediment exposure scenarios. Frequency of exceedance has not been addressed by any of the existing methodologies, but may be integrated into monitoring designs. Determination of an appropriate frequency for compliance testing of bedded sediments could be problematic if the goal of the testing is to look for current sources that may be controlled. Distinguishing new and accumulated sediment contaminants may be accomplished by sampling the suspended matter. However, this approach ignores the possibility that particle properties change when particles become bedded, and consequently, the sample may not be representative of benthic organism exposures. Another possibility is to use the freely dissolved concentration in porewater because this contaminant fraction is expected to degrade or dissipate faster than the bound fraction and thus better reflects environmental exposures. Bioavailability is discussed further in Sect. 6.1.3.

6.1.2 Multipathway Exposure

Chemicals can exist in both free and bound states within sediments. Benthic organisms may be exposed to sediment contaminants via (1) absorption of freely dissolved chemical from sediment porewater or overlying water through gills and/or body surfaces, (2) ingestion of contaminated sediment or food particles, and (3) through direct contact with sediment-bound contaminants. The predominant uptake route will be both species- and compound-dependent, and several studies have demonstrated that different exposure routes dominate uptake depending on conditions (Boese et al. 1990; Ingersoll et al. 2000; Lohmann et al. 2004; Lu et al. 2004; Savage et al. 2002; Selck et al. 2003). It is important to consider that a SQC that overlooks an exposure route is at risk of being underprotective. All possible exposure pathways should therefore be included for derivation of SQC.

The EqP approach inherently disregards all exposure pathways that do not take place via freely dissolved residues in porewater, because sediment toxicity is predicted using water-only exposures in which organisms are fed uncontaminated food. The USEPA has rationalized this approach through evidence that accumulation of hydrophobic organic compounds (HOCs) varies little across organisms of differing feeding habits, indicating that ingestion does not generally dominate exposure (Tracey and Hansen 1996). The EU EqP method accounts for potential dietary exposure for chemicals with log K_{ow} over 5 by applying an additional assessment factor of 10 to the PNEC for sediment (PNEC$_{sed}$, ECB 2003). The Dutch method employs the modified EU EqP method to account for sediment ingestion (RIVM 2001). SSTTs offer a more realistic view of sediment exposure, because benthic organisms are tested in direct contact with contaminated sediments. If the tested organism naturally feeds on sediment particles, this exposure pathway will also be accounted for by performing a SSTT. Yet, ingestion of contaminated food is overlooked in standard test methods, which may thus underestimate the importance of dietary uptake compared to field situations, where contaminant residues may also be taken up from other food sources such as algae, bacteria, and detritus.

6.1.3 Bioavailability

Contaminant bioavailability confounds sediment toxicity and must be considered when establishing any type of numerical sediment quality guideline, although bioavailability is particularly relevant for highly hydrophobic pesticides such as the pyrethroids. In the aquatic environment, nonionic organic compounds will be freely dissolved in water phases or sorbed to particulates and/or colloidal matter, which is generally referred to as dissolved organic matter (DOM). Sediment-bound contaminants represent the fraction that is sorbed to bedded and/or suspended particles and/or DOM. Many studies have indicated that the bioavailable fraction of a chemical is best represented by the freely dissolved fraction (Bondarenko et al. 2007; Bondarenko and Gan 2009; Hunter et al. 2008; Sormunen et al. 2010; Xu et al. 2007; Yang et al. 2006a, b, 2007). It is therefore widely accepted that organisms are generally not exposed to chemicals in the bound state, but rather are primarily exposed to the fraction that has desorbed and is in the freely dissolved state (You et al. 2011). Ingestion of HOCs sorbed to food particles and/or sediments has, however, also been proposed as an exposure pathway of concern, because contaminants could desorb within the organism digestive tract (Mayer et al. 2001; Mehler et al. 2011). Finally, as noted, exposure can occur through direct contact with contaminated sediment (Savage et al. 2002). For the purpose of this review, the bioavailable fraction refers to the entire chemical fraction that is available for uptake by organisms, regardless of exposure route.

Bioavailability depends on many variables including sediment characteristics (e.g., particle size, source of organic matter), organism characteristics (e.g., feeding, other behavior), chemical properties, contact time, environmental conditions (e.g., temperature, pH), and biological activity in the ecosystem (e.g., biotic transformation, cycling, and burial; Diaz and Rosenberg 1996; You et al. 2011). In this part of the review, we seek to resolve (1) the contaminant fraction(s) that organisms are at risk of being exposed to and (2) the best technique to accurately predict the bioavailable concentration for criteria derivation and compliance.

Bioavailable Fraction

As previously discussed, results from many studies have demonstrated good correlations between freely dissolved concentrations and organism uptake and/or toxicity (Bondarenko et al. 2007; Bondarenko and Gan 2009; Hunter et al. 2008; Sormunen et al. 2010; Xu et al. 2007; Yang et al. 2006a, b, 2007). Xu et al. (2007), for example, performed 10-day sediment acute toxicity tests with *Chironomus tentans*. Three sediment types were tested, and the concentrations of three pyrethroids were expressed based on five different phases: whole sediment, sediment OC, whole porewater, porewater dissolved organic carbon (DOC), and porewater freely dissolved fraction. LC_{50}s that were based on the freely dissolved fraction appeared to be matrix-independent, with little variation observed across sediments, in contrast to whole sediment and whole porewater-based LC_{50}s, which varied widely (Xu et al. 2007). Normalizing to the OC or DOC content did reduce variation in the LC_{50}s, but not entirely (Xu et al. 2007). This is one study in which the dominance of the freely dissolved contaminant fraction on organism exposure was illustrated.

Researchers have also defined the "bioaccessible" fraction. This is the fraction of chemical that, although sorbed at the time of assessment, may become available for uptake (by desorbing from sediment or DOM; Semple et al. 2004; You et al. 2011). As freely dissolved contaminant is removed from a system by organism uptake and/ or environmental transport, chemical equilibrium is disturbed, resulting in partitioning of bound contaminant to the dissolved phase until a steady state is regained. Bioavailability, which is based solely on the concentration that is currently available, may thus underestimate the total risk posed by a contaminant to benthic organisms.

Predicting or Measuring the Bioavailable Fraction

Nonionic compounds primarily sorb to OM (organic matter) contained in sediments and DOM, with the abundance of OM typically expressed as the OC content of a sorbent (Schwarzenbach et al. 2003). As discussed (Sect. 2.1), solid-water partition coefficients are often normalized to the OC content to reduce variability across sorbents. Similarly, OC normalization of sediment concentrations has been applied for various HOCs to predict the bioavailable fraction. Indeed, many studies have demonstrated good correlations between biological effects and the OC-normalized sediment concentration (Amweg et al. 2005, 2006; Trimble et al. 2008; Weston et al. 2004, 2005, 2008). Yet, this approach has several limitations that include variability in OC, effects of particle size on sorption, and sorption of highly hydrophobic compounds to mineral domains. Observed toxicity of *Hyalella azteca* in sandy sediments, for example, was lower than predicted based on measured pyrethroid concentrations in the sediments, indicating that pyrethroid bioavailability is partly controlled by the mineral phase (You et al. 2008a). In addition, it has been extensively documented that pyrethroids sorb to glassware, further supporting the mineral phase as an important sorbent for pyrethroids (Oudou and Hansen 2002; Wheelock et al. 2005; Zhou et al. 1995).

Two alternative techniques have been developed to predict or measure the freely dissolved fraction: matrix-solid-phase microextraction (matrix-SPME; Mayer et al. 2000) and Tenax® extraction (Cornelissen et al. 1997; Pignatello 1990). These techniques were thoroughly described and evaluated in a recent review by You et al.

(2011), the conclusions of which are summarized below. Both methods are matrix-independent, eliminating issues regarding variability in sediment characteristics. Matrix-SPME measures chemical activity via the freely dissolved fraction, while Tenax® addresses bioaccessibility by measuring the rapidly desorbing fraction (You et al. 2011). Despite targeting different aspects of the matrix, the two techniques have both predicted bioavailable concentrations that correlate well with biological effects and/or uptake for various pesticides and other HOCs (Parsons et al. 2007; Trimble et al. 2008; You et al. 2006, 2007, 2008a, b, 2011). Matrix-SPME is based on equilibrium partitioning and typically uses a fiber coated with poly-dimethylsiloxane as the sorbent for freely dissolved HOCs in sediment porewater. The placement of the fiber in the matrix does not disturb equilibrium because the sorption capacity of the fiber is limited to sorb less than 5% of the analytes (You et al. 2011). The residues on the fiber can be solvent extracted or thermally desorbed directly into an analytical instrument for detection and quantification of the freely dissolved chemical concentration (Mayer et al. 2000). Tenax is a sorbent suitable to measure desorption of HOCs. Tenax powder is added to a sediment-water system, altering the chemical equilibrium from the strong sorption affinity of Tenax for HOCs. As freely dissolved chemicals sorb to the Tenax, equilibrium is disrupted again, leading to desorption of more chemical from the sediment. The Tenax is subsequently removed, solvent extracted, and analyzed at different time points to measure a desorption rate. The fraction of the compound that rapidly desorbs from the sediment can be correlated to organism uptake, as this fraction is considered to represent the main source of the freely dissolved compound in porewater (You et al. 2011). An advantage of SPME is that this technique can be used in situ or in a laboratory setting (You et al. 2011). If the goal is to reach true equilibrium, SPME sampling may be too time- and labor-intensive, because it may take weeks to months for some systems to equilibrate, although one method reports sample agitation to reduce the time necessary to achieve equilibrium for the pyrethroids to less than 5 days (Hunter et al. 2009). Laboratory protocols do not require equilibrium conditions during SPME porewater extractions, only that the SPME fiber is exposed for the same length of time in each sample, typically 20–40 min (Bondarenko et al. 2007; Bondarenko and Gan 2009; Xu et al. 2007). The advantages of Tenax are the comparatively low detection limits and short processing time, relative to SPME (You et al. 2011).

Bioavailability in Current Methodologies

Some of the current SQG methodologies address bioavailability. One of the principal reasons that the USEPA selected the EqP approach for deriving sediment guidelines is that the EqP approach addresses the issue of varying bioavailability of chemicals across sediments (Di Toro et al. 2002). The EqP approach is founded on two assumptions: (1) Aquatic environments are approximately at equilibrium and (2) nonionic organic compounds sorb primarily to sediment OC or DOM (Di Toro et al. 2002). Given these assumptions, the freely dissolved concentration in porewater can be predicted from sediment concentrations using K_{oc}s, which are determined in laboratory experiments at equilibrium. Thus, predicting the bioavailable concentration in the EqP approach is highly dependent on the selected K_{oc}, which can vary greatly for the same compound. Yet, worldwide agencies advocating the EqP approach (USEPA, the Netherlands, EU, Ontario, OECD) place little to no focus on determination of site-specific sediment K_{oc}s. Such specific coefficients would seem particularly appropriate for assessment of analytically challenging compounds such as the pyrethroids, which are extremely insoluble in water. A further drawback of EqP methodologies is that these allow estimation of K_{oc} from K_{ow}, which may be less accurate than experimentally determined K_{oc}s.

All EqP approaches account for bioavailability through application of the calculated fraction of freely dissolved contaminant. Most agencies focus on bedded sediments with the exception of the EU methodology, which is based on suspended sediment. Suspended sediment contaminant concentrations are thought to reflect recent inputs of contaminants, while bedded sediments are considered repositories for sediment-associated contaminants. The EU method rationalizes the use of suspended sediment contaminant concentrations for compliance by the presumption that this fraction will settle to become the main food source for detritivorous benthic organisms (Crane 2003; Lepper 2002). For highly hydrophobic pesticides, the level of suspended solids may affect bioavailability more than partitioning to the bedded sediment. Measuring contaminant concentrations in suspended solids is one way to account for this effect, particularly in systems with high or widely fluctuating levels of suspended solids. The USEPA EqP method recognizes the dependence of bioavailability on sediment particle size distributions. The method recommends removing large particles from sediment before chemical analysis to avoid overrepresenting the contribution of large particles (with small surface areas) to sequestration of residues (Di Toro et al. 2002).

The Dutch SSTT method normalizes all SSTT data to a standard sediment (i.e., 10% OM and 25% clay on dry weight basis) to reduce the variation across different sediments used in toxicity testing (RIVM 2001). For organic chemicals, SSTT data (e.g., NOEC, LC_x, or EC_x) are normalized to the organic matter content of the standard sediment by the following equation:

$$EC_{x(\text{std sed})} = EC_{x(\text{exp})} \times H_{\text{std sed}} / H_{\text{exp}}, \tag{3}$$

where $EC_{x(\text{std sed})}$ is the experimentally derived NOEC or EC/LC_{50} normalized to standard sediment (mg/kg dry weight), $EC_{x(\text{exp})}$ is the experimentally derived NOEC or EC/LC_{50} (mg/kg dry weight), $H_{\text{std sed}}$ is the OM content (%) of the standard sediment, and H_{exp} is the OM content (%) of the experimental sediment (RIVM 2001). The OC content may be estimated from the OM content (dry weight basis) by dividing OM by a factor of 1.7, thus the OC content for standard sediment with 10% OM would make up 5.88% (RIVM 2001).

In the Canadian method, it is stated that the SSTT approach addresses bioavailability directly, because it is inherent to the method that the organisms will only be exposed to the bioavailable fraction (CCME 1995). When there is sufficient information available to define the influence of particular factors (e.g., related to bioavailability) on the toxicity of a substance, it is further required that SQGs are developed to reflect these relationships and that the relevant factors are measured in samples (CCME 1995). The Canadian method illustrates the relationship between OC content and the toxicity of nonpolar organic substances and recommends that the concentration of these substances be normalized to the sediment OC content for both SSTT and compliance testing.

A bioavailability line of evidence has been proposed as part of the weight-of-evidence approach in California (Maruya et al. 2010). This line of evidence would relate porewater concentrations, passive sampling device measurements, or measurements of the rapidly desorbing fraction of a contaminant to known toxic effects concentrations for hydrophobic organic compounds.

6.2   Summary of Methodologies

The basic methodologies, initially outlined in Sect. 2, are described in more detail here to highlight the differences between jurisdictions in the three main approaches, that is, mechanistic (EqP), spiked-sediment toxicity test, and empirical. Empirical approaches are only described in brief because most of these methods yield concentration ranges rather than single numerical values that are the focus of this review. Included in each section is a discussion on the advantages and disadvantages of each basic approach framed in terms of the ability of the approach to derive reliable numeric criteria for sediment-associated pesticides.

6.2.1   Equilibrium Partitioning (Mechanistic Approach)

The USEPA EqP Method

The EqP model was first developed by the USEPA as a mechanistic approach to SQG derivation specifically for nonionic organic compounds (Di Toro et al. 2002; USEPA 1993). The USEPA has taken this approach because it accounts for chemical bioavailability differences in varying sediment types and defines a biological effects-based concentration in sediments using available aqueous exposure data. The supporting rationale proposes that the EqP approach is likely to produce a protective concentration from biological effects in the field and that the method is sufficiently robust for use in a regulatory setting. This method utilizes the data and knowledge base built in the derivation of WQC and can serve as a tool to protect uncontaminated sites and assist in the restoration of impaired sites (Di Toro et al. 2002).

As discussed throughout the present review, the central assumption of the EqP model is that the chemical is in equilibrium between the sediment and porewater. This premise leads to the secondary assumption that the overall exposure is constant, regardless of exposure route, which is further supported by findings that exposure route has little effect on accumulation (i.e., biota-sediment accumulation factors for HOCs are similar within and among habitat groups, Tracey and Hansen 1996). When sorption to organic carbon is the main driver of bioavailability (i.e., HOCs with log $K_{ow} > 5.5$ and sediments with OC > 0.2%), the OC-normalized sediment concentration is used to predict the freely dissolved fraction from the K_{oc} (section "Bioavailability in Current Methodologies"; Di Toro et al. 2002).

Another assumption of the USEPA EqP method is that an observed effect concentration can be predicted within the uncertainty of the model (Di Toro et al. 2002). This method requires that this uncertainty is quantified to accompany the equilibrium-partitioning sediment guideline. The aspect of uncertainty involves the premise that aqueous exposure toxicity data (together with K_{oc} and OC-normalized sediment contaminant concentrations) is an appropriate predictor of sediment toxicity. This assumption has been evaluated and scrutinized over the previous decade to answer the question: Do benthic species have similar chemical exposure sensitivities as the aquatic species used to derive WQC? The USEPA has demonstrated good correlations between observed mortality and the mortality predicted using the EqP model for three sediment types and seven chemicals. Good correspondence has also been shown for acute WQC for benthic and water column organisms, supporting that benthic species (epibenthic and infaunal) do have similar sensitivities as the species used in WQC derivation (Di Toro et al. 2002).

In the USEPA EqP method, the final chronic value of the water quality criterion is considered to be representative of the chemical concentration protective of benthic life. If available, ESGs can be derived from the FCV and the K_{oc} as follows:

$$ESG = K_d \times FCV. \tag{4}$$

$$K_d = C_s / C_w = K_{oc} \times f_{oc}. \tag{5}$$

$$ESG = f_{oc} \times K_{oc} \times FCV. \tag{6}$$

$$ESG_{oc} = K_{oc} \times FCV, \tag{7}$$

where C_s is the equilibrium chemical concentration in the sediment, C_w is the freely dissolved concentration of chemical in porewater, and f_{oc} is the fraction of organic carbon in sediment (Di Toro et al. 2002). The uncertainty in the ESG can be quantified through statistical analysis as detailed by Di Toro et al. (2002). Uncertainty analysis involves data relevant to predict (1) uncertainty inherent to the method (i.e., uncertainty in applied K_{oc}s as reflected in the uncertainty of varying exposure sources, e.g., if data were derived from experiments using sediment and/or water tests or tests with varying sediment types) and (2) uncertainty related to experimental design (i.e., replication).

The USEPA has implemented a two-tiered methodology for deriving SQGs according to data availability. The minimum data requirement to compute a Tier 1 ESG is (1) the K_{ow} must be measured with current techniques, (2) the FCV should be obtained using the most current toxicity data, and (3) sediment toxicity testing must be included to validate predictions of the EqP calculation (Di Toro et al. 2002). Tier 1 ESGs are currently available for endrin, dieldrin, and PAH mixtures from the USEPA (USEPA 2003a, b, c). If data are inadequate for a Tier 1 ESG, a Tier 2 ESG may be calculated if the following requirements are met: (1) The K_{ow} is measured with current techniques and (2) the FCV or secondary chronic value is available. Sediment toxicity testing is recommended, but not required for USEPA Tier 2 ESGs (Di Toro et al. 2002).

The European Union Method

The European Union methodology is outlined in the Technical Guidance Document on Risk Assessment (ECB 2003). Data availability determines whether the EU methodology applies an assessment factor SSTT approach or the EqP approach to generate the $PNEC_{sed}$. If toxicity data are inadequate for sediment-dwelling organisms, the $PNEC_{sed}$ is calculated from the $PNEC_{water}$ and the suspended matter-water partitioning coefficient ($K_{susp-water}$) according to the EqP approach. If only acute toxicity results are available for benthic organisms (at least one datum), risk assessment is conducted on the basis of EqP predictions and the result of the most sensitive species (using an AF of 1,000). If long-term toxicity data are available for benthic organisms, the $PNEC_{sed}$ can be derived using AFs, and this is the preferred EqP method. The EU EqP method has defined the following equation for calculation of the $PNEC_{sed}$ (mg contaminant/kg sediment):

$$PNEC_{sed} = \frac{K_{susp-water}}{RHO_{susp}} \times PNEC_{water} \times 1,000, \qquad (8)$$

where $PNEC_{water}$ (mg/L) is similar to the FCV used by the USEPA, RHO_{susp} (1,150 kg/ m^3) is the bulk density of wet suspended matter, and $K_{susp-water}$ is the partition coefficient between suspended matter and water (m^3 water/m^3 suspended matter). As the EqP approach only incorporates chemical uptake via the water phase, uptake can be underestimated for compounds with log K_{ow} over 3. The level of underestimation is considered acceptable for log K_{ow} up to 5 (ECB 2003). For log K_{ow} greater than 5, as is the case with many pyrethroids, the EU EqP method is modified to account for possible sediment ingestion by reducing the calculated $PNEC_{sed}$ by an AF of 10 (ECB 2003). Further, if the measured (or predicted) chemical sediment concentration (PEC_{sed}) at a site exceeds the $PNEC_{sed}$ (i.e., $PEC_{sed}/PNEC_{sed} > 1$), then SSTTs with benthic organisms are required to refine the sediment compartment assessment (ECB 2003). As discussed (Sect. 6.1.3), the EU EqP method uses suspended solids to account for bioavailability in the SQG calculation, because suspended fluxes better reflect recent contaminant inputs as compared with bedded sediments, which are more reflective of historically deposited contaminant (Lepper 2002). The rationale for using suspended particulate matter is also applied for compliance monitoring, because newly settled particulates become the most important sediments for supporting benthic life (Lepper 2002). For compliance monitoring, the sediment quality standard is compared to the chemical concentration in the suspended matter (ECB 2003).

The Netherlands EqP Method

The Netherlands also uses the EqP approach as the basis for setting SQGs (RIVM 2001). Whereas the USEPA compute an ESG, the Dutch authorities derive the environmental risk limit for sediment. The EqP-based environmental risk limit for the sediment compartment (ERL(sed$_{EP}$)) can be calculated on dry weight basis as follows:

$$ERL(sed)_{EP} = ERL(water) \times K_d (or\ K_{oc}) \left[L/kg\ dry\ wt \right], \tag{9}$$

where ERL (water) is the environmental risk limit for aquatic species and K_d (or K_{oc}) is in units of L/kg dry wt (RIVM 2001). If experimental sediment toxicity data are available, the Dutch method requires that this data is used to evaluate if the derived ERL(sed)$_{EP}$ is sufficiently protective.

The OECD EqP Method

The OECD guidance document of aquatic effects assessment recommends implementation of either the EqP, the interstitial water quality (IWQ), or the spiked-sediment toxicity test approach for the development of SQGs (OECD 1995). The OECD EqP method describes the following equation to calculate the maximum tolerable concentration in sediment (MTC$_{sed}$):

$$MTC_{sed} = MTC_{water} \times K_d, \tag{10}$$

where MTC$_{water}$ is the maximum tolerable chemical concentration in water. Insufficient sediment toxicity data and lack of accepted sediment toxicity testing methods are described as rationales for using the EqP approach instead of the IWQ or SSTT approaches (OECD 1995). The IWQ approach is similar to EqP but applies *measured* interstitial water (porewater) concentrations in place of predicted concentrations.

The Ontario EqP Method

The Ontario guidelines recommend taking the EqP approach to determine the no-effect concentration for nonpolar organic compounds (Persaud et al. 1993). It is proposed that as this criterion corresponds to the level at which sediment contamination presents no threat to water quality and uses, benthic biota, wildlife, or human health, the no-effect level must be derived using the most stringent data (Persaud et al. 1993). The Ontario EqP method determines a SQG as

$$SQG = K_{oc} \times PWQO/G, \tag{11}$$

where PWQO/G is the provincial water quality objective/guideline. The Ontario SQG is then converted to bulk sediment concentration by assuming 1% total organic carbon sediment content.

The French EqP Method

The French system for the evaluation of quality (SEQ-Eau) implements the EqP approach to set threshold effect levels for sediment and suspended matter when toxicity data are insufficient to achieve completion of the SSTT assessment factor method (outlined in Lepper 2002). Threshold level 1 corresponds to very suitable aquatic ecosystems with negligible risk of adverse effects for all species and is calculated by application of a safety factor of 10 (e.g., NOEC/10). Threshold level 2 incorporates no safety factor and corresponds to a suitable aquatic ecosystem with possible risk of adverse chronic sublethal effects for the most sensitive species. For suspended matter, threshold levels 1 and 2 are extrapolated from the sediment thresholds, which are multiplied by a factor of 2 for organic substances.

The United Kingdom Method

The UK does not have a formal policy in place for setting SQGs; however, since 1989, the EqP approach has been applied to set values for several metals, following the method described by Pavlou and Weston (1984). The sediment action level is calculated as

$$C_{s/oc} = K_{oc} \times C_{iw}, \tag{12}$$

where $C_{s/oc}$ is the contaminant concentration in whole sediment or organic carbon and C_{iw} is the contaminant concentration in interstitial water (porewater). The $C_{s/oc}$ is equal to the sediment action level when C_{iw} equates the aquatic action level.

Evaluation of the EqP Approach

Various reports and researchers have described the advantages and disadvantages of the EqP approach. These views are compiled below with respect to the goal of developing single-value numeric SQC.

Advantages of the EqP Approach

1. EqP prediction results in a single numeric criterion.
2. WQC are available for several sediment contaminants of concern including chlorpyrifos, diazinon, and five pyrethroids.
3. The EqP approach utilizes well-established toxicological databases, eliminating the need to acquire new field data (Chapman 1989; Rowlatt et al. 2002).
4. The EqP approach incorporates organic carbon (Rowlatt et al. 2002).
5. EqP is based on chemical equilibria, which are often well known (Rowlatt et al. 2002).
6. The EqP approach efficiently identifies chemicals likely to contribute to toxicity (Rowlatt et al. 2002).
7. EqP is biologically baed to the extent that existing WQA are and thus provide more defensible guidelines than the "background approach" (Persaud et al. 1993).
8. EqP-derived SQC can be considered no-effect levels for the protection of the end uses the WQC were designed to achieve (Persaud et al. 1993).
9. EqP relies on an existing toxicological rationale (WQC), thus eliminating the need to conduct a new toxicological evaluation, as long as WQC will protect benthic organisms (Chapman 1989).

Disadvantages of the EqP Approach

1. Pesticides with log $K_{ow} > 5.5$ may confound bioavailability considerations (Di Toro et al. 2002).
2. EqP does not account for uptake through the food chain (Persaud et al. 1993; Rowlatt et al. 2002).
3. Some partition coefficients are uncertain—including those for pyrethroids and other highly hydrophobic compounds (Chapman 1989; Persaud et al. 1993; Rowlatt et al. 2002).
4. The EqP approach assumes that sediment infauna have the same sensitivity as other aquatic life (Persaud et al. 1993; Rowlatt et al. 2002).
5. With the exception of PAH mixtures, the EqP approach does not account for the presence of mixtures (Chapman 1989; Rowlatt et al. 2002).
6. The EqP approach relies on the existence of an accurate WQC for the chemical of concern (Chapman 1989; Persaud et al. 1993; Rowlatt et al. 2002).
7. Assuming equilibrium between the various phases may be unrealistic as a result of the following: (1) potentially multiphasic slow uptake kinetics for HOCs, (2) turbation of natural sediments, and (3) inability to obtain equilibrium between HOCs and growing particles (e.g., algae, plankton) from seasonal variations in particle composition, size, and association with HOCs (Crane et al. 1996; Naf et al. 1996; O'Connor and Paul 2000; Parsons et al. 2007; Rowlatt et al. 2002).
8. The EqP approach does not use toxicity data derived from the specific sediment of interest (Chapman 1989).

The prime advantage of the EqP approach is that criteria can be derived without large co-occurrence field datasets or SSTT data, both of which are limiting factors for many pesticides of concern. WQC and partitioning coefficients are, however, currently available for several pesticides making the EqP approach feasible for calculating sediment quality criteria.

The main disadvantage of the EqP approach is that some of the derived criteria have proven unrealistic because assumptions of equilibrium and partitioning to OC are not always valid. The EqP approach was pursued by the USEPA from the late 1970s to the early 1980s. In 1989, the Sediment Quality Subcommittee of the USEPA Science Advisory Board (SAB) reviewed the EqP approach and brought to light uncertainties relevant to the technical basis of the approach (USEPA 1989). The committee concluded that more information was required and that uncertainties had to be addressed before the EqP approach could be used for SQC derivation (USEPA 1989). In response, the USEPA reduced the uncertainty in the EqP approach, although as summarized in the committee's 1992 evaluation, some issues remained to be resolved (USEPA 1992).

The committee noted, for example, that limited field data prevented assessment of the uncertainty involved with extrapolation of EqP-derived values to the natural environment and that assumptions of EqP may not be valid at the particular site in question (USEPA 1992). The latter is an issue because OC is not the only factor controlling bioavailability, that is, water column species and benthic organisms may exhibit different sensitivities and short-term bioassays may underestimate long-term effects (USEPA 1992). Considerations that sediment and porewater may not be in equilibrium, that K_{ow} may not always be a good predictor of K_{oc}, and that chemical partitioning to and from sediments may be limited by kinetics further compromises the reliability and applicability of the EqP approach, although none of the above factors negate the approach (USEPA 1992).

Data for acenaphthene, dieldrin, endrin, fluoranthene, and phenanthrene supported the hypothesis that OC-normalized sediment concentrations correlate well with bioavailability; however, the supporting experiments were limited to short-term studies representing only a few species (USEPA 1992). The committee thus recommended that field experiments, including toxicity, bioaccumulation, biomarker, and/or population/community-level studies, should be conducted for method validation (USEPA 1992). The committee was also concerned that by disregarding uptake via ingestion, the EqP approach underestimates exposure, although this point has been disputed by the authors of the EqP approach (e.g., Di Toro et al. 2002). Finally, the committee acknowledged improvement of uncertainty reduction through more accurate determinations of K_{ow} but recommended that uncertainties should still be investigated and quantified (USEPA 1992).

Criticisms of the EqP approach in the literature tend to reflect the SAB review. Iannuzzi et al. (1995) argued in an editorial letter that there is consensus among scientists that sediment quality criteria should be derived via a biological effects-based approach. The authors reflect concern for the use of EqP as a regulatory tool because EqP is not directly based on biological effects, does not account for mixtures, or elucidate cause of toxicity. Instead, a co-occurrence approach was proposed to better relate to biological effects (Iannuzzi et al. 1995). A rebuttal to this letter highlights frequent misconceptions of the intended use of EqP for regulatory purposes, as well as misunderstandings related to the technical assumptions of the EqP model (Ankley et al. 1996). The rebuttal further states that the EqP model is based on valid science, that the limitations are known, and that the uncertainty in the model can be quantified, but agrees that it is inappropriate as a stand-alone tool for evaluating sediment quality on a pass/fail basis. The USEPA's use of porewater concentrations from laboratory-spiked-sediment toxicity tests for validating the EqP model has also received critique arguing that these laboratory tests may not be reflective of the environment. O'Connor and Paul (2000) explained, for example, that environmental porewater concentration data are limited and that chemicals spiked to sediments in the laboratory may exhibit unrealistically high bioavailability compared to residues aged in the field.

6.2.2 Spiked-Sediment Toxicity Test Approaches

SSTT methodologies can be carried out via several procedures. SQC can be derived from SSTT data by (1) applying an assessment factor to the lowest datum or (2) fitting all available data to a statistical distribution and define a percentile as the SQC.

SSTT Assessment Factor Approaches

As noted by TenBrook et al. (2009), all WQC AF methodologies apply the most sensitive species as a foundation for AF-based criteria and disregard factors such as taxon or toxicant mode of action; they also pool plant and animal data in the dataset. The same principles apply to SQC AF methods, wherein no discussion of pooling taxa in any of the reviewed methodologies exists. Separate criteria are, however, typically derived for freshwater and saltwater.

The European Union SSTT AF Method

The EU only employs the EqP approach as a screening tool (ECB 2003). When long-term sediment toxicity data are available, it is required that these are used with AFs to derive $PNEC_{sed}$, while a combined AF and EqP approach is accepted when only acute data are available. The EU recommends careful evaluation of studies to select the most appropriate AF for each guideline. Long-term tests with sublethal endpoints (reproduction, growth, emergence, sediment burrowing activity, and avoidance) are regarded as the most relevant exposure scenarios for sediment con-taminants (ECB 2003). Applied studies should characterize all possible exposure routes to avoid underestimation due to study design (e.g., if organisms are fed with unspiked food, exposure via sediment ingestion may be reduced). According to the EU methodology, AFs should decrease as data uncertainty decreases, with AFs designed to account for uncertainties in extrapolation of laboratory data to benthic ecosystems from intra- and inter-laboratory variation, intra- and interspecies biological variability, extrapolating from acute to chronic toxicity, and from laboratory to field impact (ECB 2003; Lepper 2002). A factor of 1–10 is added for each layer of extrapolation up to AFs of 1,000, depending on data type and quantity (ECB 2003). The $PNEC_{sed}$ is then derived from the lowest available $NOEC/EC_{10}$, obtained in long-term tests by application of selected AFs. For acute data, with the EqP method, an AF of 1,000 is applied for estimation of $PNEC_{sed}$ (detailed in Sect. 6.2.1). For one long-term datum, an AF of 100 is applied to calculate the $PNEC_{sed}$, while AFs of 50 or 10 are used if data are available for two or three long-term tests, respectively, with species representing different living and feeding conditions.

The Netherlands SSTT AF Method

For chronic datasets that are inadequate to fulfill the minimum requirement for statistical distribution analysis (i.e., representing less than four taxonomic groups, see SSD section below), the Dutch method requires completing an AF-based preliminary risk assessment (RIVM 2001). AFs range from 1 to 1,000, depending on data quantity and type (acute vs. chronic, taxa). When data requirements allow, the Dutch method refers to the EU AF method outlined above. For smaller datasets, RIVM advocates use of modified AFs according to the USEPA Great Lakes Program (USEPA 2003d).

The Canadian SSTT AF Method

The Canadian method includes a SSTT approach to be used in combination with an empirical approach (CCME 1995). If the minimum data requirements are met (i.e., four data including at least two chronic tests covering both partial and full life cycles), an assessment (safety) factor is applied to compensate for uncertainty in the dataset. The uncertainties of concern to CCME include intra- and interspecies variations, use of various endpoints, acute to chronic extrapolation, bioavailability issues, and extrapolation from laboratory to field conditions including mixtures (CCME 1995). AFs are determined on a case-by-case basis and are dependent on data availability. The SSTT approach inherently accounts for bioavailability issues and, by inclusion of relatively sensitive species in the minimum data requirements, also for interspecies variation. To account for the remaining uncertainty factors, CCME examined individual studies to quantify the margin of safety. Ratios were calculated for conversion from less sensitive test results to more sensitive test results, and ratios were combined to yield final AFs to extrapolate SQGs from acute (LC_{50}) or chronic (NOEC) data (CCME 1995). The AFs recommended by CCME are an assessment factor of 20 for acute data, while an AF of 5 is sufficient for guideline development based on chronic studies (CCME 1995). It should be noted that these assessment factors were derived based on a limited dataset for only three compounds—zinc, cadmium, and fluoranthene. The Canadian methodology proposes that different AFs could be calculated for individual chemicals or classes of chemicals or that generic AFs could be calculated for all relevant chemicals.

The French SSTT AF Method

The French methodology applies AFs ranging from 1 to 1,000 to single-species toxicity data (Lepper 2002). The size of the AF depends on the type of data available with a lower AF applied for chronic than for acute data (Lepper 2002). As discussed, threshold levels 1 and 2 correspond to the concentrations that are associated with negligible and possible risk, respectively (Sect. 6.2.1). A sediment threshold level 1 is determined from the lowest reliable NOEC or EC_{10} divided by an AF of 10 or from the lowest reliable LC/EC_{50} divided by an AF of 1,000. Threshold level 2 is set as the lowest reliable chronic NOEC (AF of 1) or the lowest reliable LC/EC_{50} divided by an AF of 100. To derive threshold levels for suspended sediment, the threshold levels for sediment are multiplied by a factor of 2.

SSTT Species Sensitivity Distribution (SSD) Approaches

Many jurisdictions apply statistical extrapolation methods to derive water quality criteria (TenBrook et al. 2009). It has been proposed that these techniques may also be suitable for deriving sediment quality criteria (e.g., ECB 2003; RIVM 2001). In SSD approaches, the cumulative probabilities of the toxicity values for a given chemical are plotted to fit a statistical distribution to the data. The criterion is derived as a percentile of the statistical distribution, typically the 5th percentile. Because of concerns of low data availability, most reviewed methodologies include little discussion on using statistical distributions for deriving SQGs. Only the Dutch and the EU methodologies specifically discuss the use of SSDs for derivation of SQGs. Both methods restrict distribution analysis to chronic data and use the 5th percentile of the distribution as the cutoff. The Dutch and EU protocols are detailed in the following.

The Netherlands SSTT SSD Method

The Dutch method recommends that an environmental risk limit (e.g., the maximum permissible concentration) be derived using a statistical approach, if sufficient data are available (i.e., chronic data for four or more species of at least four different taxonomic groups for a particular environmental compartment; RIVM 2001). The Dutch method applies a log-normal distribution to calculate the hazard concentrations (HC_p) for the 5th and 50th percentiles $(HC_5$ and $HC_{50})$:

$$\log HC_p = \bar{x} - k \times s, \tag{13}$$

where HC_p is the hazardous concentration for $p\%$ of species, $\bar{x}$ is the mean of log-transformed NOEC data, k is the extrapolation constant dependent on the percentile level of certainty and sample size (Table 1 in Aldenberg and Jaworska 2000), and s is the standard deviation of log-transformed data (RIVM 2001). Computer software has been designed for derivation of HC_ps with associated 90% confidence intervals (RIVM 2004). The HC_5 is used to set the MPC, which is used to derive environmental quality standards (RIVM 2001). The target value is the negligible concentration, which is calculated as MPC/100. In representing the distribution, where 50% of species are adversely affected, the HC_{50} indicates that an ecosystem is seriously threatened and is therefore used as an intervention value.

As of 2001, Dutch Environmental Quality Standards had been derived for 147 organic substances and pesticides in sediment; however, as the data requirement of four chronic NOECs was not met for any of these substances, the standards were derived using equilibrium partitioning (Sijm et al. 2001).

The EU SSTT SSD Method

The EU recommends priority of a statistical distribution approach if sediment toxicity data are adequate to fit to a distribution (ECB 2003). The EU has implemented the Dutch SSD procedure. However, the EU method allows deviation from log-normal distributions by recommending use of the statistical distribution that provides the best fit. The Anderson-Darling test and the Kolmogorov-Smirnov tests are proscribed methods to check goodness of fit. The PNEC is calculated for the best fit as

$$PNEC = 5\% \ SSD \ (50\%c.i.)/AF, \qquad (14)$$

where 5% SSD is the 5th percentile of the species sensitivity distribution, 50% c.i. is the 50% confidence interval, and AF ranges from 1 to 5 (ECB 2003).

The AF is determined based on professional judgment and designed to reflect additional uncertainties in the calculation. ECB recommends starting with an AF of 5, which is only reduced after consideration of (1) the overall quality of the applied database and endpoints, (2) the taxonomic diversity and representativeness of the database (e.g., are different life forms, feeding strategies and trophic levels well represented), (3) information on presumed mode of action, (4) statistical uncertainties of the 5th percentile estimate (e.g., goodness of fit, confidence intervals), and (5) comparisons of field and mesocosm data and the 5th percentile to evaluate the laboratory to field extrapolation (ECB 2003).

The EU further recommends that NOECs below the 5th percentile be discussed in the final report to assess, for example, if all these belong to the same trophic group, as this could indicate that this organism group is particularly sensitive and that some assumptions for SSD are not met (ECB 2003). The report should include a comparison of the SSD-derived and AF-derived PNECs, and the discussion should be sufficiently thorough to justify the final choice of PNEC (ECB 2003).

Evaluation of SSTT Approaches

The advantages and disadvantages of the SSTT approach are compiled below, with respect to the goal of developing single-value numeric SQC.

Advantages of the SSTT Approach

1. The SSTT approach is similar to WQC and is technically acceptable and legally defensible (Chapman 1989; Ingersoll and MacDonald 2002; Rowlatt et al. 2002).
2. SSTT does not require prior knowledge of mechanisms of uptake (Chapman 1989; Rowlatt et al. 2002).
3. SSTT allows determination of direct cause-effect relationship (Ingersoll and MacDonald 2002; Persaud et al. 1993).
4. Laboratory SSTTs can be performed with any chemical and do not require a priori assumptions about the specific mechanism of interaction or exposure route (Chapman 1989; Ingersoll and MacDonald 2002).
5. SSTTs incorporate the bioavailable contaminant fraction (Ingersoll and MacDonald 2002).

Disadvantages of the SSTT Approach

1. Few chronic data are available and only exist for a few standard test species (Chapman 1989; Rowlatt et al. 2002).
2. Little information is currently available to establish relationships between acute and chronic effects that can be used when chronic data are unavailable (Chapman 1989; Rowlatt et al. 2002).
3. It is not practical to test all possible mixtures that could occur in the environment (Chapman 1989; Rowlatt et al. 2002).
4. Presently, no basis exists for extrapolating to no-effect concentrations in sedimentary communities (Rowlatt et al. 2002).
5. Sediment-spiking techniques are not standardized, and differences in methods can strongly influence the results (Chapman 1989; Ingersoll and MacDonald 2002; Persaud et al. 1993; Rowlatt et al. 2002).
6. Laboratory tests may not be representative of field-contaminated sediment, where conditions (particularly bioavailability) may vary considerably from those in the lab (Ingersoll and MacDonald 2002; Persaud et al. 1993; Rowlatt et al. 2002).

The primary advantages of the SSTT approach are that it is technically acceptable, demonstrates direct cause-effect relationships, and addresses the issue of bioavailability. The primary disadvantages of the SSTT approach are the dearth of available data and that these data may be incompatible due to use of differing spiking methods, which can strongly influence the results.

6.2.3 Empirical Approaches

Empirical methods dominate the application of biological effects-based approaches over mechanistic methodologies. Empirical methods utilize large datasets with matching sediment chemistry and toxicity data from field-collected sediments.

Although field sediment toxicity testing indirectly accounts for mixture effects, empirical approaches do not allow causality to be attributed to any one given chemical mixture or single chemical. Empirically derived SQGs are therefore generally not intended for regulatory use, but rather for risk assessment and ranking of sites for further study. Despite these limitations, empirical approaches are sensible because large amounts of sediment chemistry and toxicity data are available through national and statewide programs. In North America, most jurisdictions utilize biological effects data, while many European countries have historically employed a reference site contamination approach to evaluate sediment contamination (Badut et al. 2005). In the Belgian province of Flanders, for example, sediment quality guidelines are based on the geometric mean of contaminant concentrations at the average reference site in combination with biological assessments and toxicity testing. A similar reference site approach has been implemented in Italy (Badut et al. 2005).

Effects Range Method (US, Canada, Australia, and New Zealand)

The effects range method was developed to interpret monitoring data collected through the NOAA National Status and Trends Program (Long and Morgan 1990; Long and MacDonald 1992; Long et al. 1995). The NSTP database of co-occurrence sediment chemistry and biological effects data was analyzed to identify the effects range low (ERL; the lower 10th percentile of the data) and the effects range median (ERM; the 50th data percentile). The dataset was modified over time to exclude freshwater data and include data from coastal and estuarine waters of the Atlantic, Pacific, and Gulf of Mexico and also to include data from equilibrium-partitioning modeling, spiked-sediment toxicity testing, and benthic community assessments (Long et al. 1995). The NOAA SQGs were derived on a sediment dry wt basis, without OC normalization, and do not directly account for bioavailability. The NOAA methodology did not aim to derive numeric biological effects-based criteria for individual contaminants, but was implemented as part of a weight-of-evidence approach to SQGs.

The NSTP database has been employed as the major data source for deriving empirically based SQGs by some non-US regulatory agencies including Australia/New Zealand (ANZECC and ARMCANZ 2000) and Canada (CCME 1995). ERL/ERM values of these jurisdictions are initially based on North American databases and are refined with local information as it becomes available. The Canadian method proscribes calculation of separate freshwater and saltwater SQGs and specifies that sediment concentrations are reported on dry wt basis (CCME 1995). If data are limited, an interim sediment quality guideline (ISQG) may be calculated, but it is required that data gaps are reported to make limitations of the ISQG transparent. The Canadian guidelines also propose for future methodologies to incorporate SSTTs to validate empirically based effect ranges.

The NOAA, USEPA, and the US Army Corps of Engineers have not adopted the NSTP empirically based SQGs, as these values were not intended for use as standards and have not been proven reliable for predicting sediment toxicity (O'Connor 2004). Thomas O'Connor of NOAA has pointed to the misinterpretation of ERL by some researchers who have incorrectly applied the ERL as the sediment chemical threshold concentration above which the probability of toxicity increases abruptly (O'Connor 2004). Rather than a toxicity threshold, "below which sediment toxicity is impossible and above which it is likely," "an ERL is simply a low point on a continuum of bulk chemical concentrations in sediment that roughly relates to sediment toxicity" (O'Connor 2004). Similar to other empirical methods, the effects range method does not allow determining the cause of toxicity and lacks consideration of chemical and environmental factors that may contribute to or be responsible for the observed biological effects.

Effects Level Method (Florida)

As part of a weight-of-evidence approach, Florida regulators have adapted the NOAA NSTP method for use in the effects level method (outlined in MacDonald et al. 1996). The Florida method includes area-specific data collected from Florida coastal waters and the Gulf of Mexico to develop a relevant biological effects database for sediment (BEDS). The effects level method is similar to the effects range method but separates biological effects and no-effect data within the database. No-effect data comprises data for which no adverse effect was observed or where an effect was observed but the average chemical concentration was less than twice the reference site level (MacDonald et al. 1996). The threshold effects level (TEL) is calculated as the geometric mean of the lower 15th percentile concentration of the effects dataset and the 50th percentile of the no-effect dataset, while the probable effects level (PEL) is defined as the geometric mean of the 50th percentile concentration of the effects dataset and the 85th percentile of the no-effect dataset (MacDonald et al. 1996). As in the effects range method, the effects level method does not account for contaminant bioavailability, which is necessary for a valid technical approach.

Apparent Effects Thresholds Method (Washington and Oregon)

The apparent effects threshold (AET) method for generating SQGs is very similar to the NSTP method but employs data specific to Oregon and Washington State. The AET method was first developed using a saltwater database of matching sediment chemistry and biological effects data from the Puget Sound, Washington (outlined in Barrick et al. 1988). The freshwater sediment quality database (FSEDQUAL) was developed across data from 33 freshwater studies and 245 stations in Washington and Oregon (Willamette River) using sediment chemistry and bioassay data from Microtox (bioluminescence inhibition in bioluminescent bacteria), *Hyalella azteca*, *Chironomus tentans*, *Daphnia magna*, *Ceriodaphnia dubia*, and *Hexagenia limbata* assays (Cubbage et al. 1997). The AET was defined by Barrick et al. (1988) as the chemical concentration above which statistically significant biological effects are always expected to occur. When Cubbage et al. (1997) developed the freshwater AET method, they developed a new SQG called the probable apparent effects threshold (PAET), which is defined as the 95th percentile of all data with no significant biological effects and no concentrations above the lowest concentration associated with effects. The AET method includes benthic community endpoints, whereas the effects range and level methods predominantly use lethality as end-points (Batley et al. 2005).

The USEPA Science Advisory Board reviewed the AET method and recommended that the method is inappropriate for developing "broadly applicable sediment quality criteria" (USEPA 1989). Concerns noted by the SAB include the method's site-specific nature, lack of validation, and inability to establish causality and to account for differences in bioavailability across sediment types. The SAB did acknowledge that the AET method is a technically valid approach for estimation of sediment quality at specific locations if appropriately validated (USEPA 1989).

Screening-Level Concentration Method (USEPA, Ontario)

The screening-level concentration method was developed from data for nonionic organic compounds (Neff et al. 1986; Persaud et al. 1993). Sediment chemistry was matched with the presence or absence of benthic species in field samples to derive species screening-level concentrations. Each SSLC was derived as the 90th percentile of the frequency distribution for a particular species at several sampling stations plotted against the OC-normalized sediment concentration for a given chemical. The SLC is then given by the 95th percentile of the frequency distribution of all SSLCs, representing an estimate of the highest contaminant concentration that can be tolerated by approximately 95% of benthic species (Neff et al. 1986).

Logistic Regression Model Method (California)

Empirically based narrative sediment objectives are currently used in California for the protection of benthic communities in bays and estuaries. Sediment toxicity, benthic community condition, and sediment chemistry data are used together to support a multiple lines of evidence approach to meet the narrative objectives. The above three metrics are used in combination, as no individual line of evidence alone is considered reliable (SWRCB 2011). Sediment toxicity and benthic community condition categories are combined in a decision matrix to determine the severity of biological effects, while sediment toxicity and chemistry categories are used to assess the potential for chemically mediated effects. Sampling stations are categorized as unimpacted, likely unimpacted, possibly impacted, likely impacted, clearly impacted, or inconclusive, with the statuses of unimpacted and likely unimpacted considered to meet the protection objectives (SWRCB 2011). A station level assessment is conducted through a decision matrix that considers the potential for chemically mediated effects and the severity of effects using the Chemical Score Index and the California Logistic Regression Model.

Logistic regression models can be used to relate a chemical's concentration in sediment to the probability of toxic effects (Field et al. 1999, 2002). Matching sediment chemistry and toxicity data from marine amphipods were initially compiled in a database for deriving the logistic regression equation used for SQG formulation in California (Field et al. 1999, 2002; summarized in Wenning et al. 2005). SWRCB highlighted several limitations of the logistic modeling method, including the possibility of over- or underestimating risk to benthic communities, that causality cannot be attributed to specific chemicals, and the lack of consideration of particle size, physical disturbance, or organic enrichment as possible contributors to biological effects (SWRCB 2011). Ritter et al. (2012) compared the performance of different empirical approaches to deriving SQGs, including an effects range median, a logistic regression model, a sediment quality guideline quotient 1 (SQGQ1), and a consensus approach. A new benthos-based chemical score index method was also compared to these existing empirical methods and found to be of the most consistently high performance, although overall results indicated that all methods performed similarly (Ritter et al. (2012)).

Probable Effects Concentration Method (Great Lakes, USA)

A probable effects concentration method was implemented to assess sediment quality in the Great Lakes region. PECs were derived across matching chemistry and toxicity data from 92 reports and over 1,600 field-collected sediment samples from throughout North America (Ingersoll et al. 2001). The applied toxicity data include results of 10- to 14- and 28- to 42-day sediment tests with the amphipod *Hyalella azteca* and 10- to 14-day tests with the midges *Chironomus riparius* and *C. tentans*. The PECs derived proved to be good predictors of sediment toxicity on both a regional and national basis (Ingersoll et al. 2001). Furthermore, mean PEC quotients (i.e., the concentration of a chemical divided by its PEC) were applied as overall measures of chemical contamination that support evaluation of sediment-contaminant mixture effects (Ingersoll et al. 2001). The PEC method is considered a consensus-based approach to sediment quality guidelines.

Evaluation of Empirical Approaches

For several reasons, empirical approaches are generally not suitable for deriving single-value numeric criteria for pesticides. The primary disadvantage of empirical approaches is their inability to resolve cause and magnitude of effect for individual compounds when multiple contaminants are present in the field samples. The possibility that co-occurring contaminants and/or other factors than the chemical of concern contributed to the observed effects cannot be ruled out, and as such, empirical methods do not provide reliable numeric criteria for individual compounds.

Furthermore, empirical approaches do not account for bioavailability variation across sediment types and thus lack a valid technical foundation for deriving reliably predictive guidelines. Empirical approaches are further limited by the requirement of site-specific datasets of co-occurring biological and chemical data, which are unavailable for many current-use pesticides and areas of interest. Some empirical methods require or recommend performing additional field bioassays and benthic community surveys, which are economically and practically unfeasible for many projects.

A recent evaluation of the probable effects, effects range, and logistic range model approaches demonstrated that national data predictions for all tested methods underestimated sediment toxicity (10-day tests) for the Calcasieu Estuary, Louisiana (MacDonald et al. 2011). A PEC model derived using Calcasieu Estuary data achieved more reliable predictions, however, highlighting the importance of developing site-specific approaches for risk assessment of benthic organisms (MacDonald et al. 2011). Fuchsman and Barber (2000) compared predictions of co-occurrence data with results of SSTT and brought attention to data for two organochlorine compounds, for which concentrations greatly exceeding empirically predicted toxic levels caused no adverse effects in SSTT. These findings demonstrate the risk of attributing toxicity of co-occurring contaminants to the incorrect chemical (Fuchsman and Barber 2000).

6.3 Important Additional Considerations for SQC Derivation

6.3.1 Mixtures

Decade-long field monitoring corroborates the ubiquitous presence of contaminant mixtures in environmental sediments (Gilliom 2007). Comprehensive protection of organisms encountering chemical mixtures in the field requires mixture effects to be incorporated into methods for criteria derivation and compliance. The EqP and SSTT approaches do not easily integrate mixture toxicity into criteria calculation. Empirical methods inherently account for mixtures by screening sediment samples for multiple compounds and performing field sediment bioassays, which detect the combined toxicity of all contaminants present, whether detected by chemical analysis or not. Yet, empirical approaches do not provide a direct link between the observed toxicity and any individual chemical or fully characterized fraction of chemicals within the mixture. This lack of defined concentration-response relationships complicates the development of a management plan.

Chemical mixtures act via additivity, synergism, and/or antagonism. The current state of knowledge on pesticide interactions is inadequate to predict the joint action of pesticide mixtures when the pesticides involve different pesticide classes that have different modes of action (Lydy et al. 2004). The concentration addition model predicts the additive toxicity of chemicals that exhibit the same mode of action, but do not interact with each other (Plackett and Hewlett 1952). The concentration addition model is the best established of the models for joint chemical toxicity and is widely used, for example, in the Water Quality Control Plan (Basin Plan) for the Sacramento River and San Joaquin River basins, in cases where multiple chemicals with similar modes of action are present in a water body (CRWQCB-CVR 2006). The concentration addition concept is also considered in the California SQG methodology (Maruya et al. 2010).

Toxic unit (TU) analysis is often applied to predict additive toxicity for chemical mixtures (Altenburger et al. 2000). A toxic unit is calculated by dividing the measured concentration in sediment (C_{sed}) by the laboratory-derived LC_{50} for a particular organism (Lydy et al. 2004):

$$TU = C_{sed}/LC_{50}. \tag{15}$$

Typically, both the sediment concentration and LC_{50} will be normalized to the organic carbon content of the respective sediments. Weston et al. (2004) used OC-normalized TUs for pyrethroids and organochlorines to identify sediment contaminants present at sufficiently high concentrations to cause the observed toxicity. The CDPR has also employed TU analysis with OC normalization to predict the potential toxicity of pyrethroids in sediments to the amphipod *Hyalella azteca* (Starner and Kelley 2005). In this assessment, the CDPR assumed additivity for all pyrethroids and simply summed the TUs for all pyrethroids present in a sample. A sum of TUs above 1 indicates that the sediment contaminant concentration exceeds the LC_{50}, enabling prediction of the expected toxicity by proportionality. The toxic unit approach is a simple tool to determine additive toxicity (TenBrook et al. 2010) that can be used in compliance by replacing the LC_{50} with the SQC of the respective compound. The application of TUs for mixture toxicity assessment is discussed in relation to measurement of bioavailable fractions for the bioavailability line of evidence in the California SQG methodology (Maruya et al. 2010).

Nonadditive mixture effects were addressed with regard to criteria compliance by TenBrook et al. (2010). They concluded that at this time, nonadditive mixture effects can only be considered in criteria compliance when valid interaction coefficients (K) are available over a range of concentrations. When a synergist or an antagonist is present in a mixture with the pesticide of interest, an interaction coefficient (K) can be used to describe the joint mixture effect (Finney 1942; Mu and LeBlanc 2004):

$$K_x = EC_{50(0)}/EC_{50(x)}, \tag{16}$$

where K_x is the interaction coefficient when the synergist/antagonist is at concentration x, $EC_{50(0)}$ is the EC_{50} determined for a chemical in absence of the synergist/antagonist, and $EC_{50(x)}$ is the EC_{50} determined for a chemical in the presence of a synergist/antagonist at concentration x.

When K_x is known, the measured concentration of a pesticide can be adjusted to account for mixture effects:

$$C_a = C_m \times K_x, \tag{17}$$

where C_a is the adjusted concentration of a pesticide (e.g., synergism ($K_x > 1$) results in concentration increase), C_m is the measured concentration of the pesticide, and K_x is the interaction coefficient for a synergist/antagonist at concentration x (Mu and LeBlanc 2004).

For use in criteria compliance, K_xs must be known for a wide range of synergist/antagonist concentrations to enable establishment of concentration-mixture effect relationships, from which the adjusted pesticide concentration can be predicted for any given synergist/antagonist concentration. These relationships need to be established for multiple species or for one particularly sensitive species to compare the adjusted concentrations to the criteria for compliance determination (TenBrook et al. 2010).

The USEPA derived SQGs for a group of PAHs through a model combining EqP, QSARs, TUs, additivity, and concentration-response models to address the mixture effects of PAHs in sediments (Swartz et al. 1995). The PAH mixture model was generated using data acquired from both published and unpublished sources of spiked-sediment toxicity tests and field sediment bioassays that followed standard methods (Swartz et al. 1995). A more general approach to PAH mixtures and narcotic chemical mixtures was developed by Di Toro and McGrath (2000) and has been officially issued by the USEPA as a method to derive equilibrium sediment benchmarks for PAH mixtures (USEPA 2003c).

To summarize, additive mixture effects can generally be incorporated in criteria compliance using the concentration addition model, which is applicable to the EqP and SSTT approaches when it has been established that it is reasonable to assume additivity. In such cases, toxic unit analysis is a simple method to check for compliance, provided that there is reliable single compound LC_{50} data available for each compound in the mixture. For nonadditive mixtures, interaction coefficients are useful if ample data are available. More complex mixtures, involving both synergists and antagonists, cannot be incorporated in compliance determination at this time, although some complex models do exist to predict such mixture effects (Rider and LeBlanc 2005; TenBrook et al. 2009). As discussed throughout this section, empirical SQG methodologies inherently incorporate mixture toxicity but are unable to attribute toxic effects to any particular chemical or chemical mixture.

6.3.2   Threatened and Endangered Species

Threatened and endangered species (TES) are particularly sensitive to stressors in the environment, and it should be ensured that sediment quality criteria will be protective of these species. As TES typically inhabit limited ranges, assessments should be restricted to local species. No existing SQG methodology specifically addresses protection of TES, perhaps because it is assumed that TES will be protected if the most sensitive species in an ecosystem are protected, as is the basis for current approaches. With few SSTT data available in general, toxicity data for TES are likely to be even more scarce and the ability to predict threshold values for TES would be very valuable. Unfortunately, no interspecies correlations have been developed for sediment toxicity based on surrogate species. Existing QSARs were developed for narcotic compounds and are unlikely to be appropriate candidates for many current-use pesticides, which often act via specific modes of action. Because of limited data availability, it may be unfeasible at this time to develop specific procedures for protection of benthic TES; however, any available data should be utilized to evaluate the ability of derived criteria to adequately protect listed species.

6.3.3 Bioaccumulation/Secondary Poisoning

Chemicals with potential for accumulation in sediments often also have a propensity to accumulate in organisms, which may lead to secondary poisoning effects as contaminants biomagnify through the food chain. According to the OECD, substances with a log $K_{ow} > 3$ and molecular weight $< 1,000$ have potential to bioaccumulate (OECD 1995). These properties are similar to those described for chemicals likely to accumulate in sediments. Several SQG methodologies address bioaccumulation by estimating the expected contaminant level in predators of aquatic organisms at the sediment concentration equal to the SQG (CCME 1995; ECB 2003; RIVM 2001). For the EqP approach, it is most appropriate to assess bioaccumulation potential via the water quality criteria used to derive to the SQG, as this removes one layer of extrapolation. Guidance for extrapolating secondary poisoning effects from aqueous concentrations is covered by TenBrook et al. (2010).

The Canadian methodology defines bioaccumulation in the context of SQGs as "the process by which substances are accumulated by aquatic organisms from all routes of exposure" (CCME 1995). To address bioaccumulation up the food chain and protect higher trophic levels, the method allows SQGs to be calculated from bioaccumulation factors (BAFs) and tissue residue guidelines for the protection of wildlife consumers of aquatic life. If the tissue-based SQG is lower than the ecotoxicity-based SQG, the final SQG may be altered to be protective of all trophic levels (CCME 1995).

The Dutch methodology includes consideration of secondary poisoning of predators caused by uptake of substances in prey organisms (RIVM 2001). According to this method, biota-sediment accumulation factors (BSAFs) and chronic toxicity studies for birds and mammals are used to assess possible secondary poisoning effects. The individual bird or mammal NOECs are divided by a BSAF to predict the corresponding sediment concentration, which can be included in the ecotoxicity dataset and used for deriving the environmental risk limit. A separate MPC is also calculated for predators and is compared to the MPC calculated for the sediment compartment (RIVM 2001).

The EU method recommends that a chemical is assessed for bioaccumulation if there is no mitigating property (e.g., hydrolysis half-life less than 12 h) and the compound matches any of the following characteristics: "A log $K_{ow} \geq 3$, is highly adsorptive, belongs to a class of substances known to have a potential to accumulate in living organisms, there are indications from structural features" (ECB 2003). For compounds with log $K_{ow} \geq 4.5$, direct uptake of freely dissolved chemicals may be of less importance than ingestion of contaminated food or sediment (ECB 2003). The EU provides guidance on assessing risk to predators from dietary uptake based with the assumption that uptake of freely dissolved contaminants is the main exposure route for prey species, but this guidance does not include exposure from contaminated sediments. Risk to a predator is assessed with two assumptions, (i) aqueous uptake is the only exposure route for prey and (ii) prey species inhabit a water body of a contaminant level equal to the predicted environmental concentration in water (PEC_{water}) as follows (ECB 2003):

$$PEC_{oral, predator} = PEC_{water} \times BCF_{fish} \times BMF_{fish}, \tag{18}$$

where $PEC_{oral, predator}$ is the predicted environmental concentration a predator will receive in the prey (diet), BCF_{fish} is the bioconcentration factor for fish on a wet wt basis (concentration per fish wet weight/concentration in water), and BMF_{fish} is the biomagnification factor in fish on lipid wt basis (concentration per fish lipid weight/concentration per diet lipid weight) (ECB 2003). The BCF_{fish} describes water column uptake only, while the BMF_{fish} represents contaminant uptake via ingestion of lower trophic level organisms.

To assess the risk of secondary poisoning from sediment contaminants, the PEC_{water} can be replaced with the PEC_{sed}:

$$PEC_{oral, predator} = PEC_{sed} \times BCF_{fish} \times BMF_{fish}. \tag{19}$$

If $NOEC_{oral, predator}$, BCF_{fish}, and BMF_{fish} are known, PECs can be substituted with NOECs to solve for $NOEC_{sediment}$:

$$NOEC_{sediment} = NOEC_{oral, predator} / BCF_{fish} \times BMF_{fish}. \tag{20}$$

If measured BCF or BMF values are not available, the EU methodology provides default values for organic substances based on the log K_{ow} of the substance. As BAF encompasses both dietary and aqueous uptake, measured BAFs can be used in place of $BCF_{fish} \times BMF_{fish}$. The $NOEC_{sediment}$ can be compared to the SQG to assess if secondary poisoning is expected for sediment concentrations equal to the SQG.

The Australia/New Zealand methodology provides specific guidance to verify that a derived SQG does not lead to contaminant levels in aquatic organisms that exceed human health standards for consumption of seafood (ANZECC and ARMCANZ 2000). A separate sediment criterion can be calculated based on the bioconcentration factor and sediment-related factors (as derived in van der Kooij et al. 1991):

$$C_{sed} = C_{org} \times K_d / r \times BCF, \tag{21}$$

where C_{sed} is the BCF-based sediment criterion, C_{org} is the human health standard for fish, r is an empirical concentration ratio of suspended matter to sediment ($r=2$

for organics, van der Kooij et al. 1991), K_d may be calculated from K_{oc}, and the applied BCF should represent a sediment-ingesting aquatic organism. The BCF-based criterion can be compared to the derived SQG to ensure that this will be protective of human health standards (ANZECC and ARMCANZ 2000).

To summarize, secondary poisoning is a crucial consideration in SQC derivation because the physicochemical properties associated with sediment accumulation are similar to those associated with bioaccumulation. Several techniques have been developed to predict toxicity due to bioaccumulation and biomagnification, and many methodologies have already incorporated these procedures into the criteria derivation process. Utilization of BCFs, BAFs, and BMFs, along with dietary exposure toxicity data and/or human health standards, is useful to ensure that a proposed SQC will not lead to secondary poisoning.

6.3.4   Harmonization Across Media

Harmonizing a sediment quality criterion across various media ensures that the derived SQC will not cause excursions of criteria into other compartments. Sediment contaminants may desorb into porewater or overlying water and aqueous contaminants may volatilize into the air, which may in turn deposit onto soil or water surfaces. Due to this close association between sediment and overlying water, harmonization across these particular compartments is the focus in most SQG methodologies that address this topic.

Cross-media harmonization is intrinsic to the EqP approach, through which SQGs are derived based on the chronic water quality criterion (FCV) and the sediment OC-water equilibrium-partitioning coefficient (Di Toro et al. 2002). In this way, a system at equilibrium with a sediment contaminant concentration equal to or below the SQG will also exhibit an aqueous contaminant concentration equal to or below the WQC. All EqP methodologies (USEPA, OECD, the Netherlands, EU, France, UK, and Ontario) thus harmonize across the sediment and water compartments but overlook other environmental compartments, such as air and soil.

The Dutch method recommends that the maximum permissible concentration derived for the primary (emission) compartment of concern be used to predict concentrations in receiving compartments at steady state via the EqP model (RIVM 2001). If the predicted concentrations exceed the MPCs derived for these (receiving) compartments (i.e., the MPC of the emission compartment leads to MPC exceedances in the receiving compartments), the MPC of the emission compartment needs adjustment. When the emission compartment is sediment, the harmonization procedure depends on the method used to derive the MPC_{sed}. If MPC_{sed} is determined by EqP, the obtained value is inherently harmonized with the water compartment and no further assessment is needed. If data are sufficient to use a statistical extrapolation to determine MPC_{sed}, no further harmonization is required because this is considered a refined effects assessment. If the sediment environmental risk limit is derived using assessment factors, this value should be compared to the value derived using the EqP approach and the lower of the two should be applied as the harmonized environmental risk limit (RIVM 2001).

In conclusion, it is sensible to assess harmonization between the sediment and water compartments to avoid criteria excursions across media, and it is likely that WQC are in place for compounds for which SQC are being developed. If EqP is used to derive SQC, the sediment and water compartments are inherently harmonized, whereas SQC derived using assessment factors or statistical distributions need to be harmonized with the aqueous compartment.

6.3.5   Utilization of all Available Data and Encouragement to Generate Data

The overall scarcity of ecotoxicity data available for benthic species constitutes a major obstruction for deriving sediment quality criteria of adequate statistical certainty. As noted for WQC by TenBrook et al. (2009), the uncertainty associated with developed criteria limits the ability of regulators to assess if a criterion is at risk of over- or underprotecting targeted species. TenBrook et al. proposed that the selected method should encourage the generation of additional ecotoxicity data by all stakeholders, because comprehensive high-quality toxicity datasets best enable producing criteria with a high level of certainty. Of the three main approaches discussed in the present review, only the empirical approach utilizes all available data to derive SQGs. The EqP approach essentially disregards any existing sediment toxicity data in favor of (likely) larger aquatic toxicity datasets, and the assessment factor SSTT approach only utilizes the single lowest value in the dataset. The SSD SSTT approach employs all available laboratory toxicity data but excludes field bioassays. Yet, the empirical approaches are not able to generate single compound numeric criteria based on causal relationships between sediment chemistry and observed biological effects; thus, we are left with the EqP and SSTT approaches.

If the EqP approach is implemented, stakeholders have no incentive to generate sediment toxicity data, because this approach only uses aquatic data, which tend to be more abundant than sediment data. Adoption of the SSTT approaches, on the other hand, would encourage data generation by either the AF or SSD procedures because the magnitude of AFs decreases and the certainty of SSDs increases as the number of data increase. Thus, to utilize all available data and encourage data generation, it may be appropriate to incorporate flexible derivation procedures into a SQC methodology, depending on the size of the dataset. If data for five or more taxa are available, for example, it would be sensible to apply an SSD, while an AF approach is appropriate for 1–5 taxa, in which a decreasing AF are applied as data increases. If no sediment data are available, the EqP approach is acceptable. SSDs allow uncertainty to be quantified, which provides information regarding the degree to which resulting criteria are likely to be under- or overprotective. For this reason, it is recommended that the SSD SSTT approach is prioritized when data requirements are met.

7 Summary

In this review, we evaluated three main current approaches for deriving sediment quality guidelines: empirical, mechanistic (equilibrium partitioning), and spiked-sediment toxicity testing approaches. Empirical approaches determine ranges of sediment concentrations that are likely or unlikely to cause toxicity, based on large datasets of matching sediment chemistry, field, and laboratory toxicity data. The empirical approaches are not suitable for determining SQC for specific pesticides because (1) direct cause-effect relationships between single sediment contaminants and toxicity cannot be discerned; (2) chemistry measurements have not accounted for bioavailability, which leads to numeric values with high uncertainty and low reliability; and (3) for many pesticides, little or no matching chemistry and toxicity data are available. In the EqP approach, SQC are derived by entering existing aquatic toxicity data into the equilibrium-partitioning model. This approach is practical for pesticides with water quality criteria in place, but the assumption of equilibrium in aquatic ecosystems is questionable, and the EqP approach neglects available sediment toxicity data. The SSTT approaches utilize sediment toxicity data, creating a scientifically defensible foundation for SQC, but experimental uncertainties regarding spiking technique and equilibration times are yet to be eliminated. The species sensitivity distribution approach generates criteria with confidence intervals, providing a measure of uncertainty, but requires relatively large datasets, whereas the assessment factor method lacks quantification of uncertainty but only requires few data to calculate conservative criteria. Several existing methodologies incorporate a combination of approaches that is dependent on data availability and the physico-chemical properties of the compound of interest.

A summary of the differences and similarities between key elements of the seven methodologies emphasized in this review is displayed in Table 6. One important element regarding sediment contamination is the incorporation of bioavailability and multiple exposure routes, which must be addressed to achieve a technically defensible methodology. It is crucial that bioavailability be incorporated in both criteria derivation and compliance determination (sampling) to ensure that data are comparable. Recent research on bioavailability of sediment contaminants has indicated that the freely dissolved porewater fraction corresponds well with uptake and toxicity. For species having significant exposure via ingestion of contaminated food and/or sediments and/or direct sediment contact, exposure may be underpredicted if these additional exposure routes are overlooked. Future SQC methodologies will be greatly improved by accounting for factors relevant for bioavailability and exposure pathways. To develop a completely new methodology, existing methodologies offer valuable building blocks that are well suited for adaptation. A new method will be more reliable and robust if it utilizes more refined risk assessments than currently are available in existing methodologies. To date, the most comprehensive methodologies for deriving single numeric SQC are those of the Netherlands and the EU, which include both SSTT and EqP approaches.

Table 6 Comparison of existing methodologies for deriving numeric sediment quality criteria

Method components			USEPA 2002[a]	RIVM 2001	ECB 2003	CCME 1995	Ontario 1993[b]	France 2002[c]	OECD 1995
Overall use of ecotoxicity data for derivation		Applies data directly		x	x	x		x	
		Applies data indirectly	x	x	x		x	x	x
		Nontraditional endpoints			x	x			
Equilibrium Partitioning (EqP)	Applied K_{oc}	Measured	x	x	x		x	x	x
		Estimated	x	x	x		x	x	x
		Site-specific							
	Derivation of FCV	Species sensitivity distribution (SSD)	x	x	x	x	x	x	x
		Assessment factor (AF)	x	x	x				x
Spiked-sediment toxicity testing (SSTT)	SSD	Distribution stated		x	x				
		Minimum number of data required		4	10				
		Minimum number of taxa required		4	8				
	AF	Minimum number of data required		1	1	4		1	
		Minimum number of taxa required		1	1	2		1	
Factors considered in criteria derivation		Acute		x	x	x		x	x
		Chronic preferred						x	
		Magnitude	x	x	x	x	x	x	x
		Duration							
		Frequency							
		Bioavailability	x	x	x	x	x	x	x
		Bioaccumulation		x	x	x			
		Additivity	x		x				
		Threatened and endangered species							

(continued)

Table 6 (continued)

Method components			USEPA 2002[a]	RIVM 2001	ECB 2003	CCME 1995	Ontario 1993[b]	France 2002[c]	OECD 1995
Sediment factors considered in criteria derivation	Normalization of sediments	Standardized sediment		x					
		Organic carbon content	x	x	x	x	x	x	x
		Suspended sediments			x			x	
		Bedded sediments	x	x		x	x	x	x
		Ingestion		x	x				

CCME = Canadian Council of Ministers of the Environment, ECB = European Chemicals Bureau, OECD = Organisation for Economic Co-operation and Development, RIVM = The Netherlands National Institute for Public Health and the Environment, USEPA = United States Environmental Protection Agency

FCV = final chronic value, K_{oc} = organic carbon-normalized solid-water partition coefficient

[a]Di Toro et al. 2002

[b]Persaud et al. 1993

[c]Lepper 2002

Acknowledgments We would like to thank the following reviewers: Daniel McClure (CRWQCB-CVR), Xin Deng (California Department of Pesticide Regulation), Dominic Di Toro (University of Delaware), and G. Fred Lee and Anne Jones-Lee (G. Fred Lee & Associates). This project was funded through a contract with the CRWQCB-CVR. Mention of specific products, policies, or procedures does not represent endorsement by the CRWQCB-CVR.

References

Aldenberg T, Jaworska JS (2000) Uncertainty of the hazardous concentration and fraction affected for normal species sensitivity distributions. Ecotox Environ Safe 46:1–18

Altenburger R, Backhaus T, Boedeker W, Faust M, Scholze M, Grimme LH (2000) Predictability of the toxicity of multiple chemical mixtures to *Vibrio fischeri*: mixtures composed of similarly acting chemicals. Environ Toxicol Chem 19:2341–2347

Amweg EL, Weston DP, Ureda N (2005) Use and toxicity of pyrethroid pesticides in the Central Valley, CA, USA. Environ Sci Technol 24:966–972

Amweg EL, Weston DP, You J, Lydy MJ (2006) Pyrethroid insecticides and sediment toxicity in urban creeks from California and Tennessee. Environ Sci Technol 40:1700–1706

Ankley GT, Berry WJ, Di Toro DM, Hansen DJ, Hoke RA, Mount DR, Reiley MC, Swartz RC, Zarba CS (1996) Use of equilibrium partitioning to establish sediment quality criteria for non-ionic chemicals: a reply to Iannuzzi et al. Environ Toxicol Chem 15:1019–1024

ANZECC, ARMCANZ (2000) Australian and New Zealand guidelines for fresh and marine water quality. Australian and New Zealand Environment and Conservation Council and Agriculture and Resource management Council of Australia and New Zealand, Canberra, Australia

Apitz SE, Power EA (2002) From risk to sediment management: an international perspective. J Soil Sediment 2:61–66

ASTM (1990) Standard guide for conducting 10-day static sediment toxicity tests with marine and estuarine amphipods. American Society of Testing and Materials. ASTM designation: E1367–90. p 24

ASTM (1990) Standard guide for conducting sediment toxicity tests with freshwater invertebrates. American Society of Testing and Materials. ASTM designation: E1383–90

ASTM (2004) Standard guide for conducting *Daphnia magna* life-cycle toxicity tests. American Society of Testing and Materials. ASTM designation: E1193–97 (2004)

ASTM (2006) Standard guide for conducting laboratory toxicity tests with freshwater mussels. American Society of Testing and Materials. ASTM designation: E2455–06

ASTM (2006) Standard guide for conducting three-brood, renewal toxicity tests with *Ceriodaphnia dubia*. American Society of Testing and Materials. ASTM designation: E1295–01 (2006)

ASTM (2007) Standard guide for conducting sediment toxicity tests with polychaetous annelids. American Society of Testing and Materials. ASTM designation: E1611–00 (reapproved 2007)

ASTM (2007) Standard guide for conducting whole sediment toxicity tests with amphibians. American Society of Testing and Materials. ASTM designation: E2591–07

ASTM (2008) Standard tests method for measuring the toxicity of sediment-associated contaminants with freshwater invertebrates. American Society of Testing and Materials. ASTM designation: E1706–05 (2008)

ASTM (2008) Standard tests method for measuring the toxicity of sediment-associated contaminants with estuarine and marine invertebrates. American Society of Testing and Materials. ASTM designation: E1367–03 (2008)

ASTM (2008) Standard guide for collection, storage, characterization, and manipulation of sediments for toxicological testing and for selection of samplers used to collect benthic invertebrates. American Society of Testing and Materials. Report E1391–03

ASTM (2008) Standard guide for designing biological tests with sediments. American Society of Testing and Materials. ASTM designation: E1525–02 (2008)

ASTM (2010) Standard guide for determination of the bioaccumulation of sediment-associated contaminants by benthic invertebrates. American Society of Testing and Materials. ASTM designation: 1688–10

Badut MP, Ahlf W, Batley GE, Camusso M, de Deckere E, den Besten PJ (2005) International overview of sediment quality guidelines and their uses. In: Wenning RJ, Batley GE, Ingersoll CG, Moore DW (eds) Use of sediment quality guidelines and related tools for the assessment of contaminated sediments. SETAC Press, Pensacola, FL, pp 345–381

Bailey HC, Deanovic L, Reves E, Kimball T, Larson K, Cortright K, Connor V, Hinton DE (2000) Diazinon and chlorpyrifos in urban waterways in Northern California, USA. Environ Toxicol Chem 19:82–87

Balthis WL, Hyland JL, Fulton MH, Pennington PL, Cooksey C, Key PB, DeLorenzo ME, Wirth EF (2010) Effects of chemically spiked sediments on estuarine benthic communities: a controlled mesocosm study. Environ Monit Assess 161:191–203

Barrick R, Becker S, Brown L, Beller H, Pastorok R (1988) Sediment quality values refinement: volume 1—1988 update and evaluation of Puget Sound AET. PTI Environmental Services, Bellevue, WA, p 74

Batley GE, Stahl RG, Badut MP, Bott TL, Clark JR, Field LJ, Ho KT, Mount DR, Swartz RC, Tessier A (2005) Scientific underpinnings of sediment quality guidelines. In: Wenning RJ, Batley GE, Ingersoll CG, Moore DW (eds) Use of sediment quality guidelines and related tools for the assessment of contaminated sediments. SETAC Press, Pensacola, FL, pp 39–119

Bockting GJM, Van de Plassche EJ, Struijs J, Canton JH (1993) Soil-water partition coefficients for organic compounds. Report 679101 013, RIVM, Bilthoven, The Netherlands

Boese BL, Lee H II, Specht DT, Randall RC (1990) Comparison of aqueous and solid-phase uptake for hexachlorobenzene in the tellinid clam *Macoma nasuta* (Conrad): a mass balance approach. Environ Toxicol Chem 9:221–231

Boethling RS, MacKay D (2000) Handbook of property estimation methods for chemicals. Environmental and health sciences. Lewis Publishers, Boca Raton, FL

Bondarenko S, Gan J (2009) Simultaneous measurement of free and total concentrations of hydrophobic compounds. Environ Sci Technol 43:3772–3777

Bondarenko S, Spurlock F, Gan J (2007) Analysis of pyrethroids in sediment pore water by solid-phase microextraction. Environ Toxicol Chem 26:2587–2593

Burton GA Jr (2002) Sediment quality criteria in use around the world. Limnology 3:65–75

Burton GA, Ingersoll CG, Burnett LC, Henry M, Hinman ML, Klaine SJ, Landrum PF, Ross P, Tuchman M (1996) A comparison of sediment toxicity test methods at three Great Lake areas of concern. J Great Lakes Res 22:495–511

California SWRCB (2011) State Water Resources Control Board web site. Available from: http://www.waterboards.ca.gov/about_us/water_boards_structure/mission.shtml. Accessed 4 Jun 2012

Caquet T, Hanson ML, Roucaute M, Graham DW, Lagadic L (2007) Influence of isolation on the recovery of pond mesocosms from the application of an insecticide. II. Benthic macroinvertebrate responses. Environ Toxicol Chem 26:1280–1290

CCME (1995) Protocol for the derivation of Canadian sediment quality guidelines for the protection of aquatic life. Canadian Council of Ministers of the Environment. CCME EPC-98E

CDPR Pesticide Data Index. Available at: http://apps.cdpr.ca.gov/ereglib/main.cfm. Accessed 4 Jun 2012

Chapman PM (1989) Current approaches to developing sediment quality criteria. Environ Toxicol Chem 8:589–599

Chapman PM (2000) Editorial: why are we still emphasizing chemical screening-level numbers? Mar Pollut Bull 40:465–466

Chapman PM (2007) Editorial: do not disregard the benthos in sediment quality assessments! Mar Pollut Bull 54:633–635

Chapman PM, Mann GS (1999) Sediment quality values (SQVs) and ecological risk assessment (ERA). Mar Pollut Bull 38:339–344

Clean Water Act (2002) Federal water pollution control act. As amended through Public Law 107-303, 2002. Available at: http://cfpub.epa.gov/npdes/cwa.cfm?program_id=45. Accessed 4 Jun 2012

Conrad AU, Fleming RJ, Crane M (1999) Laboratory and field response of *Chironomus riparius* to a pyrethroid insecticide. Water Res 33:1603–1610

Cornelissen G, van Noort PCM, Govers HAJ (1997) Desorption kinetics of chlorobenzenes, polycyclic aromatic hydrocarbons, and polychlorinated biphenyls: sediment extraction with Tenax® and effects of contact time and solute hydrophobicity. Environ Toxicol Chem 16:1351–1357

Crane M (2003) Proposed development of sediment quality guidelines under the European Water Framework Directive: a critique. Toxicol Lett 142:195–206

Crane M, Everts J, van de Guchte F, Heimbach F, Hill I, Matthiessen P, Stronkhorst J (1996) Research needs in sediment bioassay and toxicity testing. In: Munawar M, Dave G (eds) Development and progress in sediment quality assessment: rationale, challenges, techniques and strategies. SPB Academic Publishing, Amsterdam, The Netherlands

CRWQCB-CVR (2006) Proposed 2006 CWA Section 303(d) List of water quality limited segments. Central Valley Regional Water Quality Control Board web site. Available from: http://www.waterboards.ca.gov/water_issues/programs/tmdl/docs/303dlists2006/swrcb/r5_final303dlist.pdf. Accessed 4 Jun 2012

CRWQCB-CVR (2009) The water quality control plan (basin plan) for the California Regional Water Quality Control Board Central Valley Region, fourth edition, the Sacramento river basin and the San Joaquin river basin. Available from: http://www.swrcb.ca.gov/rwqcb5/water_issues/basin_plans/sacsjr.pdf. Accessed 4 Jun 2012

Ctgb (2012) Ctgb pesticides database. Board for the Authorisation of Plant Protection Products and Biocides (Ctgb), Wageningen, The Netherlands. Available at: http://www.ctb.agro.nl/por tal/page?_pageid=33,47131&_dad=portal&_schema=PORTAL. Accessed 4 Jun 2012

Cubbage J, Batts D, Breindenbach S (1997) Creation and analysis of freshwater sediment quality values in Washington State. Washington State Department of Ecology, Environmental Investigations and Laboratory Service Program, Olympia, WA

Day KE, Clements WH, DeWitt T, Landis WG, Landrum P, Morrisey DJ, Riley M, Rosenberg DM, Suter GW (1995) Workgroup summary report on critical issues of ecological relevance in sediment risk assessment, chapter 12. In: Ingersoll CG, Dillon T, Biddinger GR (eds) Ecological risk assessment of contaminated sediments. SETAC Press, Pensacola, FL

de Bruijn J, Busser F, Seinen W, Hermens J (1989) Determination of octanol/water partition coefficients for hydrophobic organic chemicals with the slow stirring method. Environ Toxicol Chem 8:499–512

Diaz RJ, Rosenberg R (1996) The influence of sediment quality on functional aspects of marine benthic communities. In: Munawar M, Dave G (eds) Development and progress in sediment quality assessment: Rationale, challenges, techniques and strategies. SPB Academic Publishing, Amsterdam, The Netherlands

Dileanis PD, Brown DL, Knifong DL, Saleh D (2003) Occurrence and transport of diazinon in the Sacramento River and selected tributaries, California, during two winter storms, Jan–Feb 2001. United States geological survey, water-resources investigations report 03–4111

Di Toro DM, McGrath JA (2000) Technical basis for narcotic chemicals and polycyclic aromatic hydrocarbon criteria. II. Mixtures and sediments. Environ Toxicol Chem 9:1487–1502

Di Toro DM, Hansen DJ, DeRosa LD, Berry WJ, Bell HE, Reiley MC, Zarba CS (2002) Technical basis for the derivation of equilibrium partitioning sediment quality guidelines (ESGs) for the protection of benthic organisms: nonionic organics. Draft report 822-R-02-041, USEPA, Office of Science and Technology and Office of Research and Development, Washington, DC

Domalgalski JL, Weston DP, Zhang M, Hladik M (2010) Pyrethroid insecticide concentrations and toxicity in streambed sediments and loads in surface water of the San Joaquin Valley, California, USA. Environ Toxicol Chem 29:813–823

ECB (2003) Technical guidance document on risk assessment in support of commission directive 93/67/EEC on risk assessment for new notified substances, commission regulation (EC) no. 1488/94 on risk assessment for existing substances, directive 98/8 EC of the European Parliament and of the council concerning the placement of biocidal products on the market. Part II. Environmental Risk Assessment. European Chemicals Bureau, European Commission Joint Research Center, European Communities

Environment Canada (1992) Biological test method: acute test for sediment toxicity using marine and estuarine amphipods. Report EPS 1/RM/26, Environmental protection, conservation and protection, Ottawa, Canada

Environment Canada (1992) Toxicity tests using luminescent bacteria (*Photobacterium phosphoreum*). Report EPS 1/RM/24, Environmental protection, conservation and protection, Ottawa, Canada

Environment Canada (1992) Fertilization assay with echinoids (sea urchins and sand dollars). Report EPS 1/RM/27, Environmental protection, conservation and protection, Ottawa, Canada

Environment Canada (1994) Guidance for the collection, handling, transport, storage and manipulation of sediments for chemical characterization and toxicity testing. Draft report, Environmental protection, conservation and protection, Ottawa, Canada

Environment Canada (1995) Guidance document on measurement of toxicity test precision using control sediments spiked with a reference toxicant. Report EPS 1/RM/30, Ottawa, Canada.

Environment Canada (1997) Biological test method: test for growth and survival in sediment using larvae of freshwater midges (*Chironomus tentans* or *Chironomus riparius*). Technical report EPS 1/RM/32, Ottawa, Canada

Environment Canada (1997) Biological test method: test for growth and survival in sediment using the freshwater amphipod *Hyalella azteca*. Technical report EPS 1/RM/33, Ottawa, Canada

Field LJ, MacDonald DD, Norton SB, Ingersoll CG, Severn CG, Smorong D, Lindskoog R (2002) Predicting amphipod toxicity from sediment chemistry using logistic regression models. Environ Toxicol Chem 21:1993–2005

Field LJ, MacDonald DD, Norton SB, Severn CG, Ingersoll CG (1999) Evaluating sediment chemistry and toxicity data using logistic regression modeling. Environ Toxicol Chem 18: 1311–1322

Finney DJ (1942) The analysis of toxicity tests on mixtures of poisons. Ann Appl Biol 29:82–94

Fleming RJ, van de Guchte C, Grootelaar L, Ciarelli S, Borchert J, Looise B, Guerra MT (1996) Sediment tests for poorly water soluble substances: European intra- and inter-laboratory comparisons. In: Munawar M, Dave G (eds) Development and progress in sediment quality assessment: rationale, challenges, techniques and strategies. SPB Academic Publishing, Amsterdam, The Netherlands

Forbes VE, Cold A (2005) Effects of the pyrethroid esfenvalerate on life-cycle traits and population dynamics of *Chironomus riparius* - Importance of exposure scenario. Environ Toxicol Chem 24:78–86

Fuchsman PC, Barber TR (2000) Spiked sediment toxicity testing of hydrophobic organic chemicals: bioavailability, technical considerations, and applications. J Soil Contam 9:197–218

Gerstl Z (1990) Estimation of organic chemical sorption by soils. J Contam Hydrol 6:357–375

Gilliom RJ (2007) Pesticides in U.S. streams and groundwater. Environ Sci Technol 41:3409–3414

Hansch C, Leo A, Hoekman D (1995) Exploring QSAR. Hydrophobic, electronic, and steric constants. American Chemical Society, Washington, DC

Hatakeyama S, Yokoyama N (1997) Correlation between overall pesticide effects monitored by shrimp mortality test and change in macrobenthic fauna in a river. Ecotoxicol Environ Saf 36:148–161

Hladik ML, Kuivila KM (2009) Assessing the occurrence and distribution of pyrethroids in water and suspended sediments. J Agric Food Chem 57:9079–9085

Holmes RW (2004) Monitoring of sediment-bound contaminants in the lower Sacramento River watershed. Surface water ambient monitoring program (SWAMP). Lower Sacramento River watershed. Final report. California Environmental Protection Agency. Regional Water Quality Control Board Central Valley Region. Available at: http://www.waterboards.ca.gov/centralvalley/water_issues/water_quality_studies/sedimentchem.pdf. Accessed 4 Jun 2012

Hose GC, Lim RP, Hyne RV, Pablo F (2002) A pulse of endosulfan-contaminated sediment affects macroinvertebrates in artificial streams. Ecotoxicol Environ Saf 51:44–52

Hunter W, Xu YP, Spurlock F, Gan J (2008) Using disposable polydimethylsiloxane fibers to assess the bioavailability of permethrin in sediment. Environ Toxicol Chem 27:568–575

Hunter W, Yang Y, Reichenberg F, Mayer P, Gan J (2009) Measuring pyrethroids in sediment pore water using matrix-solid phase microextraction. Environ Toxicol Chem 28:36–43

Iannuzzi TJ, Bonnevie NL, Huntley SL, Wenning RJ, Truchon SP, Tull JD, Sheehan PJ (1995) Comments on the use of equilibrium partitioning to establish sediment quality criteria for non-ionic chemicals. Environ Toxicol Chem 14:1257–1259

Ingersoll CG, Ivey CD, Brunson EL, Hardesty DK, Kemble NE (2000) An evaluation of the toxicity: whole sediment vs. overlying-water exposures with the amphipod *Hyalella azteca*. Environ Toxicol Chem 19:2906–2910

Ingersoll CG, MacDonald DD (2002) A guidance manual to support the assessment of contaminated sediments in freshwater ecosystems, volume III—interpretation of the results of sediment quality investigations. Great Lakes National Program Office, Chicago, IL, EPA-905-B02-001-C

Ingersoll CG, MacDonald DD, Wang N, Crane JL, Field LF, Haverland PS, Kemble NE, Lindskoog RA, Severn C, Smorong DE (2001) Predictions of sediment toxicity using consensus-based freshwater sediment quality guidelines. Arch Environ Contam Toxicol 41:8–21

Kalf DF, Mensink BJWG, Montforts MHMM (1999) Protocol for the derivation of harmonized maximum permissible concentrations (MPCs). RIVM report 601506 001, National Institute of Public Health and the Environment, Bilthoven, The Netherlands

Karickhoff SW, Long, JM (1995) Internal report on summary of measured, calculated and recommended LogKow values. Internal Report, Environmental Research Laboratory, US Environmental Protection Agency, Athens, GA

Kratzer CR, Zamora C, Knifong DL (2002) Diazinon and chlorpyrifos loads in the San Joaquin River Basin, California, Jan–Feb 2000. United States geological survey, water-resources investigations report 02–4103

La Point TW, Belanger SE, Crommentuijn T, Goodrich-Mahoney J, Kent RA, Mount DI, Spry DJ, Vigerstad T, Di Toro DM, Keating FJ Jr, Reiley MC, Stubblefield WA, Adams WJ, Hodson PV, Erickson RJ, (2003) Problem formulation. In: Reiley MC, Stubblefield WA, Adams WJ, Di Toro DM, Hodson PV, Erickson RJ Keating FJ (eds) Reevaluation of the state of the science for water-quality criteria development. SETAC Press, Pensacola, FL. pp. 1–14

Laskowski DA (2002) Physical and chemical properties of pyrethroids. Rev Environ Contam Toxicol 174:49–170

Lepper P (2002) Towards the derivation of quality standards for priority substances in the context of the water framework directive. Final report of the study contract No B4-3040/2000/30637/MAR/E1: Identification of quality standards for priority substances in the field of water policy, Fraunhofer-Institute Molecular Biology and Applied Ecology, Munich, Germany

LOGKOW (2000) LOGKOW octanol-water partition coefficient program. Now KowWin. Syracuse Research Corporation, New York, NY. Available at: http://esc.syrres.com/esc/est_soft.htm. Accessed 4 Jun 2012

Lohmann R, Burgess RM, Cantwell MG, Ryba SA, MacFarlane JK, Gschwend PM (2004) Dependency of polychlorinated biphenyl and polycyclic aromatic hydrocarbon bioaccumulation in *Mya arenaria* on both water column and sediment bed chemical activities. Environ Toxicol Chem 23:2551–2562

Long JM, Karickhoff SW (1996) Protocol for setting Kow values. Draft, Sept 1996. US Environmental Protection Agency, National Exposure Research Laboratory, Ecosystems Research Division, Athens, GA, p 19

Long ER, MacDonald DD (1992) National status and trends program approach. In: Sediment classification methods compendium, chapter 14. US Environmental Protection Agency, Washington, DC. EPA 823-R-92-006

Long ER, MacDonald DD, Smith SL, Calder FD (1995) Incidence of adverse effects within ranges of chemical concentrations in marine and estuarine sediments. Environ Manage 19:81–97

Long ER, Morgan LG (1990) The potential for biological effects of sediment-sorbed contaminants tested in the national status and trends program. NOAA technical memorandum NOS OMA 52. National Oceanic and Atmospheric Administration. Seattle, Washington

Loring DH, Rantala RTT (1992) Manual for the geochemical analysis of marine sediments and suspended particulate matter. Earth-Sci Rev 32:235

Lu X, Reible DD, Fleeger JW (2004) Relative importance of ingested sediment versus pore water as uptake routes for PAHs to the deposit-feeding oligochaete *Ilyodrilus templetoni*. Arch Environ Contam Toxicol 47:207–214

Lydy M, Belden J, Wheelock C, Hammock B, Denton D (2004) Challenges in regulating pesticide mixtures. Ecol Soc 9(6):1

MacDonald DD (1994) Approach to the assessment of sediment quality in Florida coastal waters, vol 1—development and evaluation of sediment quality assessment guidelines. Florida Department of Environmental Protection, Tallahassee, FL

MacDonald DD, Carr RS, Calder FD, Long ER, Ingersoll CG (1996) Development and evaluation of sediment quality guidelines for Florida coastal waters. Ecotoxicology 5:253–278

MacDonald DD, Ingersoll CG, Berger TA (2000) Development and evaluation of consensus-based sediment quality guidelines for freshwater ecosystems. Arch Environ Contam Toxicol 39:20–31

MacDonald DD, Ingersoll CG, Smorong DE, Sinclair JA, Lindskoog R, Wang N, Severn C, Gouguet R, Meyer J, Field J (2011) Baseline ecological risk assessment of the Calcasieu Estuary, Louisiana: Part 2. An evaluation of the predictive ability of effects-based sediment-quality guidelines. Arch Environ Contam Toxicol 61:14–28

MacKay D, Shiu W-Y, Ma K-C (1999) Illustrated handbook of physical-chemical properties and environmental fate for organic chemicals. CRC-LLC netbase, CD-rom version

Maruya KA, Landrum PF, Burgess RM, Shine JP (2010) Incorporating contaminant bioavailability into sediment quality assessment frameworks. In: Stephen B, Weisberg SB, Miller K (eds) Southern California coastal water research project 2010 annual report. Southern California coastal water research project, Costa Mesa, CA, pp 153–175

Mayer P, Vaes WHJ, Wijnker F, Legierse KCHM, Kraaij RH, Tolls J, Hermens JLM (2000) Sensing dissolved sediment porewater concentrations of persistent and bioaccumulative pollutants using disposable solid-phase microextraction fibers. Environ Sci Technol 34:5177–5183

Mayer LM, Weston DP, Bock MJ (2001) Benzo[a]pyrene and zinc solubilization by digestive fluids of benthic invertebrates—a cross-phyletic study. Environ Toxicol Chem 20:1890–1900

Mehler WT, Li H, Pang J, Sun B, Lydy MJ, You J (2011) Bioavailability of hydrophobic organic contaminants in sediment with different particle-size distributions. Arch Environ Contam Toxicol 61:74–82

Mensink BJWG, Montforts M, Wijkhuizen Maslankiewiez L, Tibosh H, Linders JBHJ (1995) Manual for summarizing and evaluating the environmental aspects of pesticides. Report 679101-022 RIVM, Bilthoven, The Netherlands

Mu XY, LeBlanc GA (2004) Synergistic interaction of endocrine-disrupting chemicals: model development using an ecdysone receptor antagonist and a hormone synthesis inhibitor. Environ Toxicol Chem 23:1085–1091

Mudroch A, Azcue JM (1995) Manual of Aquatic Sediment Sampling. Lewis Publishers/CRC Press, Boca Raton, FL

Naf C, Axelman J, Broman D (1996) Organic contaminants in sediments of the Baltic Sea: distribution, behaviour and fate. In: Munawar M, Dave G (eds) Development and progress in sediment quality assessment: rationale, challenges, techniques and strategies. SPB Academic Publishing, Amsterdam, The Netherlands

Neff JM, Bean DJ, Cornaby BW, Vaga RM, Gulbransen TC, Scanlon JA (1986) Sediment quality criteria methodology validation: calculation of screening level concentrations from field data. Report to US Environmental Protection Agency, Office of Water Regulations and Standards, Washington, DC, p 60

O'Connor TP (2004) The sediment quality guideline, ERL, is not a chemical concentration at the threshold of toxicity. Mar Pollut Bull 49:383–385

O'Connor TP, Paul JF (2000) Misfit between sediment toxicity and chemistry. Mar Pollut Bull 40:59–64

OECD (1992) Guidelines for testing chemicals, section 2: effects on biotic systems. Test no. 210: fish, early-life stage toxicity test. Organisation for Economic Co-operation and Development, Paris, France

OECD (1995) Guidance document for aquatic effects assessment. Organisation for Economic Co-operation and Development, Paris, France

OECD (2004) Guidelines for testing chemicals, section 2: effects on biotic systems. Test no. 202: Daphnia sp. acute immobilization test. Organisation for Economic Co-operation and Development, Paris, France

OECD (2004) Guidelines for testing chemicals, section 2: effects on biotic systems. Test no. 218: Sediment-water chironomid toxicity using spiked sediment. Organisation for Economic Co-operation and Development, Paris, France

OECD (2004) Guidelines for testing chemicals, section 2: effects on biotic systems. Test no. 219: Sediment-water chironomid toxicity using spiked water. Organisation for Economic Co-operation and Development, Paris, France

OECD (2007) Guidelines for testing chemicals, section 2: effects on biotic systems. Test no. 225: sediment-water lumbriculus toxicity test using spiked sediment. Organisation for Economic Co-operation and Development, Paris, France

OECD (2008) Guidelines for testing chemicals, section 2: effects on biotic systems. Test no. 211: Daphnia magna reproduction test. Organisation for Economic Co-operation and Development, Paris, France

OPP Pesticide Ecotoxicity Database. Available at: http://www.ipmcenters.org/Ecotox/DataAccess. cfm. Accessed 4 Jun 2012

Oudou HC, Hansen HCB (2002) Sorption of lambda-cyhalothrin, cypermethrin, deltamethrin, and fenvalerate to quartz, corundum, kaolinite, and montmorillonite. Chemosphere 49:1285–1294

Parsons J, Jesus Belzunce Segarra M, Cornelissen G, Gustafsson O, Grotenhuis T, Harms H, Janssen CR, Kukkonen J, van Noort P, Ortega Calvo JJ, Solaun Etxeberria O (2007) Characterization of contaminants in sediments—effects of bioavailability on impact. In: Barcelo D, Petrovic M (eds) Sustainable management of sediment resources: sediment quality and impact assessment of pollutants. Elsevier, Oxford, UK

Pavlou SP, Weston DP (1984) Initial evaluation of alternatives for development of sediment related criteria for toxic contaminants in marine waters (phase II). Report prepared for US Environmental Protection Agency, Washington, DC

Persaud D, Jaagumagi R, Hayton A (1993) Guidelines for the protection of and management of aquatic sediment quality in Ontario. Ontario Ministry of the Environment, Water Resources Branch

Pignatello JJ (1990) Slowly reversible sorption of aliphatic halocarbons in soils. I. Formation of residual fractions. Environ Toxicol Chem 9:1107–1115

Plackett RL, Hewlett PS (1952) Quantal responses to mixtures of poisons. J Roy Stat Soc B 14:141–163

Raimondo S, Vivian DN, Barron MG (2010) Web-based interspecies correlation estimation (Web-ICE) for acute toxicity: user manual. Version 3.1. Office of Research and Development, US Environmental Protection Agency, Gulf Breeze, FL. EPA/600/R-10/004

Rider CV, LeBlanc GA (2005) An integrated addition and interaction model for assessing toxicity of chemical mixtures. Toxicol Sci 87:520–528

Ritter KJ, Bay SM, Smith RW, Vidal-Dorsch DE, Field LJ (2012) Development and evaluation of sediment quality guidelines based on benthic macrofauna responses. Integr Environ Assess Manag 8:610–624. doi:10.1002/ieam.191

RIVM (2001) Guidance document on deriving environmental risk limits. Traas TP (ed) RIVM report 601501 012. National Institute of Public Health and the Environment, Bilthoven, The Netherlands

RIVM (2004) ETX 2.0. A program to calculate hazardous concentrations and fraction affected, based on normally distributed toxicity data. van Vlaardingen PLA, Traas TP, Wintersen, AM and Aldenberg T. RIVM report 601501028/2004. National Institute of Public Health and the Environment (RIVM), Bilthoven, The Netherlands

RIVM (2012) RIVM library. National Institute of Public Health and the Environment, Bilthoven, The Netherlands. Available at: http://www.rivm.nl/en/Search/Library. Accessed 4 Jun 2012

Rowlatt S, Matthiessen P, Reed J, Law R, Mason C (2002) Recommendations on the development of sediment quality guidelines. Environment Agency, Bristol, UK

Savage WK, Quimby FW, DeCaprioi AP (2002) Lethal and sublethal effects on polychlorinated biphenyls on *Rana sylvatica* tadpoles. Environ Toxicol Chem 21:168–174

Schwarzenbach RP, Gschwend PM, Imboden DM (2003) Environmental organic chemistry, 2nd edn. Wiley, Hoboken, NJ

Selck H, Palmqvist A, Forbes VE (2003) Uptake, depuration, and toxicity of dissolved and sediment-bound fluoranthene in the polychaete. Capitella sp. I. Environ Toxicol Chem 22:2354–2363

Semple KT, Doick KJ, Jones KC, Burauel P, Craven A, Harms H (2004) Defining bioavailability and bioaccessibility of contaminated soil and sediment is complicated. Environ Sci Technol 38:228A–231A

Sijm D, De Bruijn J, Crommentuijn T, van Leeuwen K (2001) Environmental quality standards: endpoints or triggers for a tiered ecological effect assessment approach? Environ Toxicol Chem 20:2644–2648

Sormunen AJ, Tuikka AI, Akkanen J, Leppanen MT, Kukkonen JVK (2010) Predicting the bio-availability of sediment-associated spiked compounds by using the polyoxymethylene passive sampling and Tenax® extraction methods in sediments from three river basins in Europe. Arch Environ Contam Toxicol 58:80–90

Starner K, Kelley K (2005) Pyrethroid concentrations in surface water and bed sediment in high agricultural use regions of California. Poster presented at 26th Annual SETAC Conference, Nov 2005, Baltimore, Maryland

Stephan CE, Rogers JW (1985) Advantages of using regression analysis to calculate results of chronic toxicity tests. In: Bahner RC, Hansen DJ (eds) Aquatic toxicology and hazard assessment eighth symposium. ASTM STP 891. American Society for Testing and Materials, Philadelphia, PA, pp 328–338

Swartz RC, Schults DW, Ozretich RJ, Lamberson JO, Cole FA, DeWitt TH, Redmond MS, Ferraro SP (1995) □PAH: A model to predict the toxicity of field collected marine sediment contamination by polycyclic aromatic hydrocarbons. Environ Toxicol Chem 14:1977–1987

SWRCB (2011) Appendix A. Draft proposed amendments to the water quality control plan for enclosed bays and estuaries plan part 1: Sediment quality. Available at: http://www.swrcb.ca.gov/water_issues/programs/bptcp/sediment.shtml. Accessed 4 Jun 2012

TenBrook PL, Palumbo AJ, Fojut TL, Hann P, Karkoski J, Tjeerdema RS (2010) The University of California-Davis methodology for deriving aquatic life pesticide water quality criteria. Rev Environ Contamin Toxicol 209:1–155

TenBrook PL, Tjeerdema RS, Hann P, Karkoski J (2009) Methods for deriving pesticide aquatic life criteria. Rev Environ Contam Toxicol 199:19–110

Tracey GA, Hansen DJ (1996) Use of biota-sediment accumulation factors to assess similarity of nonionic organic chemical exposure to benthically-coupled organisms of differing tropic mode. Arch Environ Contam Toxicol 30:467–475

Trimble TA, You J, Lydy MJ (2008) Bioavailability of PCBs from field-collected sediments: application of Tenax extraction and matrix-SPME techniques. Chemosphere 71:337–344

USEPA (1980) Guidelines for deriving water quality criteria for the protection of aquatic life and its uses. Fed Reg 45:79318

USEPA (1985) Guidelines for developing numerical national water quality criteria for the protection of aquatic organisms and their uses. PB-85-227049. US Environmental Protection Agency National Technical Information Service, Springfield, VA

USEPA (1989) Evaluation of the Apparent Effects Threshold (AET) approach for assessing sediment quality. Report of the Sediment Criteria Subcommittee of the Science Advisory Board. EPA-SAB-EETFC-89-027

USEPA (1992) An SAB report: Review of sediment criteria development methodology for nonionic organic contaminants. Prepared by the Sediment Quality Subcommittee of the Ecological Processes and Effects Committee. EPA-SAB-EPEC-93-002

USEPA (1993) Technical basis for deriving sediment quality criteria for nonionic organic contaminants for the protection of benthic organisms by using equilibrium partitioning. Report. EPA-822-R-93-011. USEPA. Office of Science and Technology and Office of Research and Development, Washington, DC

USEPA (1994) Assessment guidance document: assessment and remediation of contaminated sediments program, Great Lakes National Program Office, Chicago, IL, EPA 905-R94-002

USEPA (1994) Methods for measuring the toxicity and bioaccumulation of sediment-associated contaminants with freshwater invertebrates. EPA 600-R24-024

USEPA (1996) Ecological effects test guidelines. OPPTS 850.1010 (draft). Aquatic invertebrate acute toxicity test, freshwater daphnids. USEPA Office of Prevention, Pesticides and Toxic Substances (OPPTS), EPA 712-C-96-114

USEPA (1996) Ecological effects test guidelines. OPPTS 850.1075 (draft). Fish acute toxicity test, freshwater and marine. USEPA Office of Prevention, Pesticides and Toxic Substances (OPPTS), EPA 712-C-96-118; OECD 203

USEPA (1996) Ecological effects test guidelines. OPPTS 850.1735 (draft). Whole sediment acute toxicity invertebrates, freshwater. USEPA Office of Prevention, Pesticides and Toxic Substances (OPPTS), EPA 712-C-96-354

USEPA (1996) Ecological effects test guidelines. OPPTS 850.1740 (draft). Whole sediment acute toxicity invertebrates, marine. USEPA Office of Prevention, Pesticides and Toxic Substances (OPPTS), EPA 712-C-96-355

USEPA (1996) Ecological effects test guidelines. OPPTS 850.1790 (draft). Chironomid sediment toxicity test. USEPA Office of Prevention, Pesticides and Toxic Substances (OPPTS), EPA 712-C-96-313

USEPA (1996) Ecological effects test guidelines. OPPTS 850.1800 (draft). Tadpole/sediment subchronic toxicity test. USEPA Office of Prevention, Pesticides and Toxic Substances (OPPTS), EPA 712-C-96-132

USEPA (1996) Ecological effects test guidelines. OPPTS 850.1850 (draft). Aquatic food chain transfer. USEPA Office of Prevention, Pesticides and Toxic Substances (OPPTS), EPA 712-C-96-133

USEPA (1996) Ecological effects test guidelines. OPPTS 850.1900 (draft). Generic freshwater microcosm test, laboratory. USEPA Office of Prevention, Pesticides and Toxic Substances (OPPTS), EPA 712-C-96-134

USEPA (1996) Ecological effects test guidelines. OPPTS 850.1925 (draft). Site-specific aquatic microcosm test, laboratory. USEPA Office of Prevention, Pesticides and Toxic Substances (OPPTS), EPA 712-C-96-173

USEPA (1996) Ecological effects test guidelines. OPPTS 850.1950 (draft). Field testing for aquatic organisms. USEPA Office of Prevention, Pesticides and Toxic Substances (OPPTS), EPA 712-C-96-135

USEPA (1998) Contaminated sediment management strategy (EPA 823-R-98-001). Available at: http://water.epa.gov/polwaste/sediments/cs/stratndx.cfm. Accessed 4 Jun 2012

USEPA (2000) Methods for measuring the toxicity and bioaccumulation of sediment-associated contaminants with freshwater invertebrates. Second edition. EPA 600/R-99/064, Office of Research and Development, Washington, DC

USEPA (2000) Prediction of sediment toxicity using consensus-based freshwater sediment quality guidelines. EPA 905/R-00/007, Great Lakes National Program Office, Chicago, IL

USEPA (2003) Procedures for the derivation of equilibrium partitioning sediment benchmarks (ESBs) for the protection of benthic organisms: Dieldrin. EPA 600 R 02 010, Office of Research and Development, Washington, DC

USEPA (2003) Procedures for the derivation of equilibrium partitioning sediment benchmarks (ESBs) for the protection of benthic organisms: Endrin. EPA 600 R 02 009, Office of Research and Development, Washington, DC

USEPA (2003) Procedures for the derivation of equilibrium partitioning sediment benchmarks (ESBs) for the protection of benthic organisms: PAH mixtures. EPA 600 R 02 013, Office of Research and Development, Washington, DC

USEPA (2003) Water quality guidance for the Great Lakes system. Federal Register, 40 CFR Part 132. US Environmental Protection Agency, Washington, DC

van der Kooij LA, van de Meent D, van Leeuwen CJ, Bruggeman WA (1991) Deriving quality criteria for water and sediment from the results of aquatic toxicity tests and product standards: application of the equilibrium partitioning method. Water Res 25:697–705

Verschueren K (1983) Handbook of environmental data on organic chemicals, 2nd edn. Van Nostrand Reinhold Co., New York

Verschueren K (2001) Handbook of environmental data on organic chemicals, CD-ROM, 4th edn. Wiley Interscience, New York

WAC (1995) Chapter 173-204 WAC: Sediment management standards. Washington Administrative Code (WAC). Available at: http://www.ecy.wa.gov/biblio/wac173204.html. Accessed 4 Jun 2012

Wallace JB, Lugthart GJ, Cuffney TF, Schurr GA (1989) The impact of repeated insecticidal treatments on drift and benthos of a headwater stream. Hydrobiologia 79:135–147

Wenning RJ, Batley GE, Ingersoll CG, Moore DW (eds) (2005) Use of sediment quality guidelines and related tools for the assessment of contaminated sediments. Society of Environmental Toxicology and Chemistry (SETAC), Pensacola, FL, p 815

Wenning RJ, Ingersoll CG (2002) Summary of the SETAC pellston workshop on use of sediment quality guidelines and related tools for the assessment of contaminated sediments, Fairmont, Montana, USA, 17–22 Aug 2002, Society of Environmental Toxicology and Chemistry (SETAC), Pensacola, FL

Weston DP, Holmes RW, You J, Lydy MJ (2005) Aquatic toxicity due to residential use of pyrethroid insecticides. Environ Sci Technol 39:9778–9784

Weston DP, You J, Lydy MJ (2004) Distribution and toxicity of sediment-associated pesticides in agriculture dominated water bodies of California's central valley. Environ Sci Technol 38:2752–2759

Weston DP, Zhang M, Lydy MJ (2008) Identifying the cause and source of sediment toxicity in an agriculture-influenced creek. Environ Toxicol Chem 27:953–962

Wheelock CE, Miller JL, Miller MJ, Phillips BM, Gee SJ, Tjeerdema RS, Hammock BD (2005) Influence of container adsorption upon observed pyrethroid toxicity to *Ceriodaphnia dubia* and *Hyalella azteca*. Aquat Toxicol 74:47–52

Woin P (1998) Short- and long-term effects of the pyrethroid insecticide fenvalerate on an invertebrate pond community. Ecotoxicol Environ Saf 41:137–156

Woodburn KB, Doucette WJ, Andren AW (1984) Generator column determination of octanol-water partition coefficients for selected polychlorinated biphenyl congeners. Environ Sci Technol 18:457–459

Xu Y, Spurlock F, Wang Z, Gan J (2007) Comparison of five methods for measuring sediment toxicity of hydrophobic contaminants. Environ Sci Technol 41:8394–8399

Yang WC, Gan JY, Hunter W, Spurlock F (2006a) Effect of suspended solids on bioavailability of pyrethroid insecticides. Environ Toxicol Chem 25:1585–1591

Yang WC, Hunter W, Spurlock F, Gan J (2007) Bioavailability of permethrin and cyfluthrin in surface waters with low levels of dissolved organic matter. J Environ Qual 36:1678–1685

Yang WC, Spurlock F, Liu WP, Gan JY (2006b) Inhibition of aquatic toxicity of pyrethroid insecticides by suspended sediment. Environ Toxicol Chem 25:1913–1919

Yasuno M, Fukushima S, Hasegawa J, Shioyama F, Hatakeyama S (1982) Changes in the benthic fauna and flora after application of temephos to a stream on Mt. Tsukuba. Hydrobiologia 89:205–214

You J, Harwood AD, Li H, Lydy MJ (2011) Chemical techniques for assessing bioavailability of sediment-associated contaminants: SPME versus Tenax extraction. J Environ Monit 13:792–800

You J, Landrum F, Lydy MJ (2006) Comparison of chemical approaches for assessing bioavailability of sediment-associated contaminants. Environ Sci Technol 40:6348–6353

You J, Pehkonen S, Landrum PF, Lydy MJ (2007) Desorption of hydrophobic compound from laboratory-spiked sediment measured by Tenax absorbent and matrix solid-phase microextraction. Environ Sci Technol 41:5672–5678

You J, Pehkonen S, Weston DP, Lydy MJ (2008a) Chemical availability and sediment toxicity of pyrethroid insecticides to Hyalella azteca: application to field sediment with unexpectedly low toxicity. Environ Toxicol Chem 27:2124–2130

You J, Weston DP, Lydy MJ (2008b) Quantification of pyrethroid insecticides at sub-ppb levels in sediment using matrix-dispersive accelerated solvent extraction with tandem SPE cleanup. In: Gan J, Spurlock F, Hendley P, Weston DP (eds) Synthetic pyrethroids: occurrence and behavior in aquatic environments. American Chemical Society, Washington, DC

Zabel TF, Cole S (1999) The derivation of environmental quality standards for the protection of aquatic life in the UK. J CIWEM 13:436–440

Zhou JL, Rowland S, Mantoura RF (1995) Partition of synthetic pyrethroid insecticides between dissolved and particulate phases. Water Res 29:1023–1031

Zvinavashe E, Murk AJ, Rietjens IMCM (2009) On the number of EINECS compounds that can be covered by (Q)SAR models for acute toxicity. Toxicol Lett 184:67–72

Index

Abiotic conditions, landfills & phthalates, **224:** 39 ff.

Abiotic degradation, phthalate esters, **224:** 43

Acid & base hydrolysis, phthalates (diag.), **224:** 45

Adverse effects, phthalate esters, **224:** 42

Analysis method, ptaquiloside, **224:** 76

Animal models, Bracken fern toxicity, **224:** 67

Animal test results, Bracken fern intake (table), **224:** 68

Animal toxicity, Bracken fern intake, **224:** 67

Aquatic life criteria, pesticides in sediments, **224:** 97 ff.

Aquatic life sediment quality criteria, protection defined, **224:** 112

Aquatic sediments, characterized, **224:** 98

Aquatic toxicity, pesticides, **224:** 99

Arsenate metabolism, arsenate reductases, **224:** 15

Arsenate reductases, arsenate metabolism, **224:** 15

Arsenate, description, **224:** 2

Arsenic acid derivatives, arsenic metabolism, **224:** 9

Arsenic contamination, bioremediation methods, **224:** 19-25

Arsenic cycle, global (diag.), **224:** 5

Arsenic cycle, microbial interactions, **224:** 1 ff.

Arsenic cycle, microbial transformations, **224:** 6

Arsenic detoxification, eukaryotic cells (diag.), **224:** 11

Arsenic forms, posttransductional regulation, **224:** 17

Arsenic metabolism, arsenic permeases, **224:** 16

Arsenic metabolism, methylarsines, **224:** 9

Arsenic permeases, arsenic metabolism **224:** 16

Arsenic pollution, sources, **224:** 5

Arsenic, ars operon & transcriptional regulation, **224:** 14

Arsenic, biosensors, **224:** 25

Arsenic, characteristics (table), **224:** 2

Arsenic, description & uses, **224:** 2

Arsenic, detoxification mechanism, **224:** 12

Arsenic, environmental distribution, **224:** 4

Arsenic, environmental fate, **224:** 4

Arsenic, environmental levels (table), **224:** 3

Arsenic, enzyme reduction, **224:** 7

Arsenic, measuring bioavailability, **224:** 25

Arsenic, methylation and demethylation, **224:** 8

Arsenic, microbial detoxification & transport (diag.), **224:** 9

Arsenic, microbial resistance, **224:** 6

Arsenic, microbial sensing, **224:** 13

Arsenic, microbial uptake & extrusion, **224:** 10

Arsenic, mobilization and immobilization, **224:** 10

Arsenic, the ars operon & proteins, **224:** 13

Arsenic, toxicity, **224:** 2

Arsenic-resistant microbes, examples (table), **224:** 6

Arsenic-resistant microbes, examples, **224:** 20

Arsenic-toxicity mitigation, organismal strategies, **224:** 20

Arsenite reduction, pathways (table), **224:** 8

Arsenite, oxidation processes, **224:** 7

Bioaccumulation, implication for deriving sediment quality criteria, **224:** 158

Bioavailability considerations, in deriving
     sediment quality criteria, **224:** 135
Bioavailability of arsenic, measurement,
     224: 25
Biodegradation, phthalates, **224:** 47
Biological organization level to protect,
     sediment quality criteria, **224:** 112
Bioremediation methods, arsenic
     contamination, **224:** 19-25
Bioremediation, described, **224:** 20
Biosensors, arsenic, **224:** 25
Bovine papillomavirus amplification,
     polymerase chain reaction, **224:** 83
Bovine papillomavirus diagnosis, polymerase
     chain reaction role, **224:** 85
Bovine papillomavirus role, tumor induction
     & progression, **224:** 64
Bovine papillomavirus, Bracken fern
     interaction, **224:** 65
Bovine papillomavirus, described, **224:** 64
Bovine papillomavirus, tumor progression &
     Bracken fern, **224:** 64
Bovine tumor classification, pathology,
     224: 66
Bracken & other ferns, cause of EBH
     (enzootic bovine hematuria) in cattle,
     224: 57
Bracken fern carcinogen, ptaquiloside, **224:** 57
Bracken fern effects, animals (table), **224:** 68
Bracken fern interaction, bovine
     papillomavirus, **224:** 65
Bracken fern plant, photos (illus.), **224:** 58
Bracken fern toxicity, animal models, **224:** 67
Bracken fern toxicity, animals, **224:** 67
Bracken fern, gene mutations, **224:** 79
Bracken fern, oxidative stress role, **224:** 82
California pesticides, types used, **224:** 99
Cancer diagnostic aid, ultrasonography,
     224: 82
Carcinogenic mechanism, ptaquiloside (diag.),
     224: 63
Carcinogenic mechanism, ptaquiloside,
     224: 62
Chemical behavior, phthalates, **224:** 39 ff.
Chemical mixtures, effect on deriving
     sediment quality criteria, **224:** 155
Chemical-physical properties, phthalates,
     224: 41
Chromosomal aberrations, human bladder
     cancer role, **224:** 81
Clastogen, ptaquiloside, **224:** 57
Clinical profile, EBH in cattle (table), **224:** 56
Clinical symptoms, of EBH in cattle, **224:** 55

Controlling EBH, approaches, **224:** 85
Criteria calculation, for sediment water
     quality, **224:** 131
Data analysis, setting sediment quality criteria
     for pesticides, **224:** 122
Data sources, for deriving sediment quality
     criteria, **224:** 114
Decomposition, phthalates, **224:** 43
Degradation, ptaquiloside, **224:** 77
Demethylation, of arsenic, **224:** 8
Detoxification & transport, arsenic (diag.),
     224: 9
Detoxification mechanism, arsenic, **224:** 12
Diagnosing bovine papillomavirus,
     polymerase chain reaction role, **224:** 85
Disease etiology, EBH in cattle, **224:** 57
EBH (enzootic bovine hematuria), disease in
     cattle described, **224:** 54
EBH cause, *Pteridium aquilinum*, **224:** 54
EBH in cattle, caused by bracken & other
     ferns, **224:** 57
EBH in cattle, disease etiology, **224:** 57
EBH, clinical symptoms, **224:** 55
EBH, enzootic localities and incidence,
     224: 54
EBH, in India, **224:** 55
EBH, prevention & control, **224:** 85
EBH, symptoms & clinical character (table),
     224: 56
Ecotoxicity data needs, acute vs. chronic
     exposure, **224:** 120
Ecotoxicity data quality, deriving sediment
     quality criteria, **224:** 124
Ecotoxicity data quantity needed, deriving
     sediment quality criteria, **224:** 129
Ecotoxicity data, role in deriving sediment
     quality criteria, **224:** 120
Empirical approaches, to developing sediment
     quality criteria, **224:** 107
Endangered species, implication for deriving
     sediment quality criteria, **224:** 157
Environmental aquatic toxicity, pesticides,
     224: 99
Environmental distribution, arsenic, **224:** 4
Environmental effects, ptaquiloside-induced
     enzootic bovine hematuria, **224:** 53 ff.
Environmental fate, arsenic, **224:** 4, 8-10
Environmental levels, natural arsenic (table),
     224: 3
Environmental management, arsenic cycle,
     224: 1 ff.
Environmental mobilization, arsenic, **224:** 10
Environmental source, phthalates, **224:** 40

Enzootic bovine hematuria (EBH),
    ptaquiloside-induced, **224:** 53 ff.
Enzootic localities, EBH, **224:** 54
Enzymatic reduction, arsenic, **224:** 7
Exposure duration, setting sediment quality
    criteria, **224:** 132
Exposure frequency, setting sediment quality
    criteria, **224:** 132
Exposure magnitude, setting sediment quality
    criteria, **224:** 132
Fern carcinogen, ptaquiloside, **224:** 55
Fern content, ptaquiloside, **224:** 60
Fern fronds, ptaquiloside content, **224:** 60, 76
Fern-induced tumors, ptaquiloside adducts,
    224: 80
Gene mutations, Bracken fern, **224:** 79
Genotoxic activity, ptaquiloside, **224:** 77
Genotoxicity, methylated As(III), **224:** 10
Genotoxin carcinogen, ptaquiloside, **224:** 57
Global cycle, arsenic (diag.), **224:** 5
Health effects, phthalates, **224:** 40
Human bladder cancer, chromosomal
    aberration role, **224:** 81
Human effects, ptaquiloside-induced enzootic
    bovine hematuria, **224:** 53 ff.
Human food chain, flow of ptaquiloside
    (diag.), **224:** 75
Human food chain, ptaquiloside entry points
    (diag.), **224:** 74
Human health effects, ptaquiloside, **224:** 73
Hydrolysis, phthalate esters, **224:** 44
Hydrolysis, phthalates transformation (diag.),
    224: 45
Immunohistochemical expression, tumor
    biomarkers, **224:** 80
India, EBH disease in cattle, **224:** 55
Landfills, phthalate behavior, **224:** 39 ff.
Landfills, source of phthalate esters, **224:** 42
Mechanism, of arsenic detoxification (diag.),
    224: 11
Mechanistic approach, to developing sediment
    quality guidelines, **224:** 107
Methodology components, sediment quality
    criteria, **224:** 101
Methylarsines, arsenic metabolism, **224:** 9
Methylated As(III), genotoxicity, **224:** 10
Methylation & demethylation, role in arsenic
    transport, **224:** 9
Methylation, of arsenic, **224:** 8
Microbes, resistant to arsenic (table), **224:** 6
Microbes, toxin sensing, **224:** 26
Microbial detoxification & transport, arsenic
    (diag.), **224:** 9

Microbial interactions, arsenic cycle, **224:** 1 ff.
Microbial oxidation, arsenite, **224:** 7
Microbial resistance, arsenic, **224:** 6
Microbial sensing, arsenic, **224:** 13
Microbial transformation, in arsenic cycle,
    224: 6
Microbial uptake, arsenic, **224:** 10
Mobilization & demobilization, arsenic,
    224: 10
Multipathway exposures, in deriving sediment
    quality criteria, **224:** 134
Mutagen, ptaquiloside, **224:** 57
Oxidation, of arsenite, **224:** 7
Oxidative stress role, Bracken fern-induced
    cancer, **224:** 82
Pathology, bovine bladder tumor classification,
    224: 66
Pesticide criteria derivation, sediments,
    224: 97 ff.
Pesticide criteria derivation, sediments,
    224: 97 ff.
Pesticide criteria for sediments, methodologies
    compared (table), **224:** 163
Pesticide toxicity, aquatic environments,
    224: 99
Pesticide types, toxicity, **224:** 99
Pesticide types, used in California, **224:** 99
Photocatalytic oxidation, phthalates, **224:** 46
Photolysis, phthalates, **224:** 47
Phthalate esters, abiotic degradation, **224:** 43
Phthalate esters, adverse effects, **224:** 42
Phthalate esters, hydrolysis, **224:** 44
Phthalate esters, in landfills, **224:** 42
Phthalate, structure (illus.), **224:** 41
Phthalates, abiotic conditions in landfills,
    224: 39 ff.
Phthalates, acid & base hydrolysis (diag.),
    224: 45
Phthalates, biodegradation, **224:** 47
Phthalates, chemical behavior, **224:** 39 ff.
Phthalates, chemical-physical properties,
    224: 41
Phthalates, degradation-decomposition, **224:** 43
Phthalates, described, **224:** 39
Phthalates, fugitive nature, **224:** 40
Phthalates, health effects, **224:** 40
Phthalates, hydrolytic transformation (diag.),
    224: 45
Phthalates, photocatalytic oxidation, **224:** 46
Phthalates, photolysis, **224:** 47
Phthalates, synthesis via esterification, **224:** 40
Physical-chemical properties, phthalates,
    224: 41

Physicochemical data needs, deriving
    sediment quality criteria, **224:** 117
Pollution sources, arsenic, **224:** 5
Polymerase chain reaction, bovine
    papillomavirus amplification, **224:** 83
Posttransductional regulation, arsenic forms,
    224: 17
Preventing EBH, approaches, **224:** 85
Ptaquilosid, carcinogenic mechanism (diag.),
    224: 63
Ptaquiloside & similars, structure (illus.),
    224: 59
Ptaquiloside adducts, fern-induced tumors,
    224: 80
Ptaquiloside, analysis method, **224:** 76
Ptaquiloside, cancer types induced, **224:** 62
Ptaquiloside, carcinogen in fern, **224:** 55
Ptaquiloside, carcinogenic mechanism,
    224: 62
Ptaquiloside, content in fern species,
    224: 57, 60
Ptaquiloside, environmental residues, **224:** 74
Ptaquiloside, fern fronds, **224:** 76
Ptaquiloside, genotoxic activity, **224:** 77
Ptaquiloside, genotoxic carcinogen, **224:** 57
Ptaquiloside, human food chain flow (diag.),
    224: 75
Ptaquiloside, human health effects, **224:** 73
Ptaquiloside, in the human food chain, **224:** 73
Ptaquiloside in human food, entry points
    (diag.), **224:** 74
Ptaquiloside, physiological behavior, **224:** 62
Ptaquiloside, soil & water effects, **224:** 75
Ptaquiloside, soil degradation, **224:** 77
Ptaquiloside, soil leaching, **224:** 76
Ptaquiloside-induced bovine hematuria,
    environmental effects, **224:** 53 ff.
Ptaquiloside-induced bovine hematuria,
    human effects, **224:** 53 ff.
Ptaquiloside-induced cancer, tumor modeling,
    224: 78
Pteridium aquilinum, EBH cause, **224:** 54
QSARs (quantitative structure activity
    relationships), in deriving sediment
    quality criteria for pesticides, **224:** 130
Reduction pathways, arsenite (table), **224:** 8
Residues in air, ptaquiloside, **224:** 74
Residues in Bracken fern, ptaquiloside,
    224: 74
Residues in water, ptaquiloside, **224:** 74
Secondary poisoning, implication for deriving
    sediment quality criteria, **224:** 158
Sediment pesticide criteria, aquatic life,
    224: 97 ff.

Sediment quality criteria derivation,
    bioaccumulation & secondary
    poisoning implication, **224:** 158
Sediment quality criteria derivation,
    bioavailability considerations,
    224: 135
Sediment quality criteria derivation,
    calculating criteria, **224:** 131
Sediment quality criteria derivation, data
    required, **224:** 114
Sediment quality criteria derivation, data
    sources, **224:** 114
Sediment quality criteria derivation,
    ecotoxicity data needs, **224:** 120
Sediment quality criteria derivation,
    ecotoxicity data needs, **224:** 120
Sediment quality criteria derivation,
    ecotoxicity data quality, **224:** 124
Sediment quality criteria derivation,
    ecotoxicity data quantity needed,
    224: 129
Sediment quality criteria derivation, endpoint
    choices, **224:** 123
Sediment quality criteria derivation, exposure
    parameters role, **224:** 132
Sediment quality criteria derivation,
    interspecies relationship data,
    224: 124
Sediment quality criteria derivation,
    methodology summary & analysis,
    224: 139-155
Sediment quality criteria derivation, mixtures
    effect, **224:** 155
Sediment quality criteria derivation,
    multipathway exposures, **224:** 134
Sediment quality criteria derivation, numeric
    criteria vs. advisory levels, **224:** 110
Sediment quality criteria derivation,
    physicochemical data needs, **224:** 117
Sediment quality criteria derivation, QSARs
    role, **224:** 130
Sediment quality criteria derivation, single- vs.
    multi-species data, **224:** 122
Sediment quality criteria derivation, species
    selected (table), **224:** 127
Sediment quality criteria derivation,
    threatened & endangered species
    implication, **224:** 157
Sediment quality criteria derivation, value of
    harmonizing across media, **224:** 160
Sediment quality criteria methodology, survey
    (table), **224:** 105-6
Sediment quality criteria, biological
    organization level to protect, **224:** 112

Sediment quality criteria, components (table), **224:** 101

Sediment quality criteria, data analysis methods, **224:** 122

Sediment quality criteria, derivation approaches, **224:** 100

Sediment quality criteria, deriving from numeric criteria vs. advisory levels, **224:** 110

Sediment quality criteria, empirical methods, **224:** 107

Sediment quality criteria, mechanistic approach, **224:** 104

Sediment quality criteria, methodologies compared (table), **224:** 163

Sediment quality criteria, methodology, **224:** 100

Sediment quality criteria, numeric criteria definitions & uses, **224:** 111

Sediment quality criteria, over- & under-protection, **224:** 113

Sediment quality criteria, portion of species to protect, **224:** 113

Sediment quality criteria, protection defined, **224:** 112

Sediment quality criteria, relevant acronyms (table), **224:** 102-4

Sediment quality criteria, spiked-sediment toxicity approach, **224:** 108

Sediment quality criteria, uses & definitions, **224:** 108

Sediment quality guidelines, mechanistic approach, **224:** 107

Sediment toxicity testing methods, selected list (table), **224:** 125

Sediments characterized, aquatic environments, **224:** 98

Soil & water effects, ptaquiloside, **224:** 75

Soil degradation, ptaquiloside, **224:** 77

Soil leaching, ptaquiloside, **224:** 76

Species selected, deriving sediment quality criteria (table), **224:** 127

Species to protect, sediment quality criteria, **224:** 113

Spiked-sediment toxicity approach, sediment quality criteria, **224:** 108

Synthesis methodology, phthalates, **224:** 40

Toxicity, arsenic, **224:** 2

Toxicity, pesticide types, **224:** 99

Toxicity, pesticides in aquatic environments, **224:** 99

Toxin sensing, microbes, **224:** 26

Tumor biomarkers, expression, **224:** 80

Tumor induction & progression, role for bovine papillomavirus, **224:** 64

Tumor modeling, ptaquiloside-induced cancer, **224:** 78

Tumorous disease of cattle, enzootic bovine hematuria, **224:** 53 ff.

Ultrasonography, cancer diagnostic aid, **224:** 82

Water & soil effects, ptaquiloside, **224:** 75